Directed Reading A

Section: Stars

Write the letter of the correct answer in the space provided.

______ **1.** What do scientists study to learn about far away stars?
  a. gravity
  b. starlight
  c. space
  d. continuous spectrum

COLOR OF STARS

______ **2.** How do scientists tell if a star is warm or cool?
  a. by its size
  b. by its age
  c. by its color
  d. by its shape

______ **3.** Which star is the coolest?
  a. a red star
  b. a white star
  c. a blue star
  d. a blue-white star

______ **4.** Which star is the warmest?
  a. a red star
  b. a yellow star
  c. a blue star
  d. an orange star

COMPOSITION OF STARS

______ **5.** What band of colors is made by white light passing through a
glass prism?
  a. a spectrum
  b. ultraviolet light
  c. a wavelength
  d. a color wheel

______ **6.** What do astronomers use to separate a star's light into a spectrum?
  a. a telescope
  b. a microscope
  c. an oscilloscope
  d. a spectroscope

| Directed Reading A *continued*

Types of Spectra

Use the terms from the following list to complete the sentences below.

continuous spectrum   emission lines

7. A spectrum that shows all the colors is called a(n)

__________________________.

8. Hot gases emit wavelengths of light, or colors, called

__________________________.

Trapping the Light: Cosmic Detective Work

Write the letter of the correct answer in the space provided.

_______ **9.** What kind of light is made by the center of a star?
    a. blue light
    b. yellow light
    c. red light
    d. white light

_______ **10.** What happens to light in the atmosphere of a star?
    a. Colors are emitted.
    b. Colors are absorbed.
    c. Colors are removed.
    d. Colors are added.

_______ **11.** What is the spectrum of a star called?
    a. an absorption spectrum
    b. a continuous spectrum
    c. an emission spectrum
    d. a nuclear spectrum

_______ **12.** What colors are shown by black lines on a star's spectrum?
    a. colors given off in the atmosphere
    b. colors absorbed by the atmosphere
    c. colors not affected by the atmosphere
    d. colors added by the atmosphere

Identifying Elements by Using Dark Lines

_______ **13.** What spectrum is used to help find elements in a star's atmosphere?
    a. an absorption spectrum
    b. a continuous spectrum
    c. an emission spectrum
    d. a nuclear spectrum

Directed Reading A *continued*

The Types of Elements in a Star

______ **14.** What elements are stars mostly made of?
    a. hydrogen and carbon
    b. helium and iron
    c. hydrogen and helium
    d. carbon and iron

______ **15.** Besides hydrogen and helium, what are the most common elements in stars?
    a. carbon, nitrogen, oxygen
    b. calcium, iron, sodium
    c. mercury, potassium, uranium
    d. chlorine, nitrogen, oxygen

CLASSIFYING STARS

______ **16.** How did early scientists group stars?
    a. by size
    b. by age
    c. by temperature
    d. by elements

Differences in Temperature

______ **17.** How do scientists now group stars?
    a. by size
    b. by age
    c. by temperature
    d. by elements

______ **18.** Which color star is the hottest?
    a. yellow   **c.** orange
    b. blue   **d.** red

Differences in Brightness

______ **19.** Before they had telescopes, what did astronomers call the brightest stars?
    a. first-magnitude stars
    b. third-magnitude stars
    c. fifth-magnitude stars
    d. sixth-magnitude stars

______ **20.** What kind of numbers are used for the magnitude of dim stars?
    a. positive numbers
    b. negative numbers
    c. whole numbers
    d. prime numbers

Directed Reading A *continued*

______ **21.** What kind of numbers are used for the magnitude of bright stars?
   a. positive numbers
   b. negative numbers
   c. whole numbers
   d. prime numbers

______ **22.** What is the magnitude of Sirius, the brightest star in the night sky?
   a. 14
   b. −1.4
   c. −9.8
   d. 9.8

HOW BRIGHT IS THAT STAR?

Apparent Magnitude

Use the terms from the following list to complete the sentences below.

   absolute magnitude   apparent magnitude

23. The brightness of a star as seen from Earth is ____________________________.

24. The brightness a star would have 32.6 light-years from Earth is

____________________________.

DISTANCE TO THE STARS

Use the terms from the following list to complete the sentences below.

   parallax   light-year

25. The distance that light travels in one year is a(n) ____________________________.

26. A star may seem to shift in place because of ____________________________.

MOTIONS OF STARS

The Apparent Motion of Stars

Write the letter of the correct answer in the space provided.

______ **27.** Why do the stars in the sky seem to circle Polaris?
   a. Earth rotates.
   b. The stars rotate.
   c. Polaris rotates.
   d. Earth stays in one place.

The Actual Motion of Stars

______ **28.** Why is it hard to see stars move?
   a. Stars move too slowly.
   b. Stars move too fast.
   c. Stars are too far away.
   d. Stars are too close.

Directed Reading A

Section: The Life Cycle of Stars

TYPES OF STARS

Write the letter of the correct answer in the space provided.

_______ **1.** Besides by mass, size, brightness, color, temperature, and composition,
how are stars classified?
  a. by constellation
  b. by age
  c. by distance from Earth
  d. by planetary system

_______ **2.** What happens to the classification of a star as its properties change?
  a. Its classification never changes.
  b. It stops being classified.
  c. Its classification changes twice.
  d. Its classification changes.

THE LIFE CYCLE OF SUNLIKE STARS

Use the terms from the following list to complete the sentences below.

_______ **3.** gravity first pulls dust and gas into a sphere

_______ **4.** the longest stage of a star's life cycle

_______ **5.** the center of a star shrinks, and its
atmosphere grows very large

_______ **6.** the leftover center of a red giant

  a. red giant
  b. white dwarf
  c. protostar
  d. main sequence

A TOOL FOR STUDYING STARS

Write the letter of the correct answer in the space provided.

_______ **7.** What diagram shows how a star's temperature and absolute magnitude
are related?
  a. the main-sequence diagram
  b. the Danish diagram
  c. the H-A diagram
  d. the H-R diagram

_______ **8.** What is one thing scientists can tell from the H-R diagram?
  a. Stars never change over time.
  b. Stars change over time.
  c. All stars are getting brighter.
  d. All stars are getting hotter.

| Directed Reading A *continued*

The H-R Diagram

_______ **9.** What appears along the bottom of the H-R diagram?
    a. temperature
    b. brightness
    c. size
    d. age

_______ **10.** What appears along the left side of the H-R diagram?
    a. temperature
    b. brightness
    c. size
    d. age

_______ **11.** Where do stars spend most of their lifetimes on the H-R diagram?
    a. lower right
    b. upper left
    c. lower left
    d. main sequence

_______ **12.** What kind of pattern is the main sequence?
    a. horizontal
    b. circular
    c. diagonal
    d. spiral

Match the correct description with the correct term. Write the letter in the space provided.

_______ **13.** top of the H-R diagram

_______ **14.** bottom of the H-R diagram

_______ **15.** right side of the H-R diagram

_______ **16.** left side of the H-R diagram

a. hot (blue) stars
b. cool (red) stars
c. dim stars
d. bright stars

Directed Reading A *continued*

THE AGING OF MASSIVE STARS

Use the terms from the following list to complete the sentences below.

> pulsar　　　　　　　　　　　　supernova
> black hole　　　　　　　　　　 neutron star

17. A huge explosion in which a large star collapses is called a(n)

_______________________.

18. A star that collapses into a small and dense ball of neutrons is a(n)

_______________________.

19. A spinning neutron star that sends out beams of radiation is a(n)

_______________________.

20. An object so massive and dense that light cannot escape its gravity is a(n)

_______________________.

Skills Worksheet

Directed Reading A

Section: Galaxies

Write the letter of the correct answer in the space provided.

______ **1.** What are large groups of stars, dust, and gas held together by gravity?
- **a.** planetary systems
- **b.** supernovas
- **c.** asteroids
- **d.** galaxies

TYPES OF GALAXIES

Match the correct description with the correct term. Write the letter in the space provided.

______ **2.** galaxy type with a bulge at the center and spiral arms

______ **3.** spiral galaxy where the sun and Earth are located

______ **4.** round or oval galaxies; called "cosmic snowballs"

______ **5.** galaxy type without a definite shape; stars form slowly

- **a.** elliptical
- **b.** irregular
- **c.** spiral
- **d.** Milky Way

CONTENTS OF GALAXIES

Gas Clouds

Write the letter of the correct answer in the space provided.

______ **6.** What is a large cloud of gas and dust where stars are born?
- **a.** spherical halo
- **b.** nebula
- **c.** cluster
- **d.** planetary system

Star Clusters

______ **7.** What is a tight group of stars that looks like a ball?
- **a.** spherical halo
- **b.** nebula
- **c.** globular cluster
- **d.** open cluster

Directed Reading A *continued*

______ **8.** What is a relatively close group of stars along the disk of a spiral galaxy?
    a. spherical halo
    b. nebula
    c. globular cluster
    d. open cluster

QUASARS

______ **9.** What is a starlike source of energy in the center of a galaxy?
    a. a nebula
    b. a quasar
    c. a cluster
    d. a Magellanic Cloud

______ **10.** What do some scientists think caused quasars?
    a. massive explosions
    b. massive black holes
    c. massive nebulas
    d. massive supernovas

ORIGIN OF GALAXIES

______ **11.** Why is looking through a telescope like looking back in time?
    a. Galaxies travel in time.
    b. Light takes time to travel.
    c. Stars travel in time.
    d. Galaxies change over time.

______ **12.** What do scientists study to learn about the early universe?
    a. nearby galaxies
    b. far galaxies
    c. spiral galaxies
    d. the sun

Directed Reading A

Section: Formation of the Universe

Write the letter of the correct answer in the space provided.

______ **1.** What is the study of the universe called?
  a. planetology
  b. cosmography
  c. astronomy
  d. cosmology

THE BIG BANG THEORY

______ **2.** What have scientists learned by studying the movement of galaxies?
  a. The universe is not moving.
  b. Galaxies are moving apart.
  c. Galaxies are moving together.
  d. The universe is getting smaller.

______ **3.** What standard model do scientists use to explain the expansion of the universe?
  a. the big bang theory
  b. the theory of universe expansion
  c. the theory of cosmology
  d. the theory of fundamental forces

A Tremendous Explosion

______ **4.** According to the big bang theory, how did the universe begin?
  a. as a cloud of gases
  b. as a sea of gases
  c. with a small explosion
  d. with a big explosion

______ **5.** According to the big bang theory, how long ago did the universe begin?
  a. about 100 billion years ago
  b. about 14 billion years ago
  c. about 10 billion years ago
  d. about 1 million years ago

Directed Reading A *continued*

Cosmic Background Radiation

_______ **6.** What is the energy in space left over from the big bang called?
　　　a. cosmic background radiation
　　　b. cosmic background energy
　　　c. space background radiation
　　　d. background radio noise

_______ **7.** Where in space did the big bang send cosmic background radiation?
　　　a. only to the solar system
　　　b. only to the closest galaxies
　　　c. in every direction
　　　d. only to the center of the universe

GRAVITY AND THE UNIVERSE

_______ **8.** After the big bang, what caused matter to form galaxies?
　　　a. random chance
　　　b. magnetism
　　　c. gravity
　　　d. nuclear fusion

_______ **9.** How does gravity control the size and shape of the universe?
　　　a. Gravity pulls galaxies together.
　　　b. Gravity pushes galaxies apart.
　　　c. Gravity makes hydrogen explode.
　　　d. Gravity makes atoms unstable.

A Cosmic Repetition

_______ **10.** What is every object in the universe a part of?
　　　a. a larger system
　　　b. a smaller system
　　　c. a solar system
　　　d. a planetary system

_______ **11.** What is one other system the Milky Way galaxy contains?
　　　a. the Andromeda galaxy
　　　b. a galaxy cluster
　　　c. our solar system
　　　d. an elliptical galaxy

Directed Reading A *continued*

HOW OLD IS THE UNIVERSE?

_______ **12.** What kind of stars are the oldest in the Milky Way galaxy?
    a. blue stars
    b. yellow stars
    c. white dwarfs
    d. protostars

_______ **13.** How old are the oldest white dwarf stars?
    a. between 12 and 13 thousand years
    b. between 12 and 13 million years
    c. between 12 and 13 billion years
    d. between 12 and 13 trillion years

A FOREVER-EXPANDING UNIVERSE?

_______ **14.** What makes up most of the total universe?
    a. matter
    b. dark matter
    c. dark energy
    d. cosmic background radiation

_______ **15.** What does dark energy seem to be doing?
    a. helping the universe get smaller
    b. helping the universe expand
    c. helping the universe stay the same size
    d. helping a new universe begin

_______ **16.** What do scientists think will happen if the universe expands forever?
    a. Stars will never age.
    b. New stars will always be born.
    c. More galaxies will form.
    d. The universe will become cold and dark.

Directed Reading B

Section: Stars

1. What is a star made of?

____

____

2. To learn about stars, astronomers study _______________________.

COLOR OF STARS

_______ **3.** What color are the hottest stars?
    a. blue
    b. yellow
    c. white
    d. red

4. What can we conclude about stars that differ in color?

____

____

COMPOSITION OF STARS

_______ **5.** The band of colors produced when white light passes through a prism
is a(n)
    a. color wheel.
    b. emission line.
    c. ultraviolet light.
    d. spectrum.

_______ **6.** A hot, solid object gives off a(n)
    a. continuous spectrum.
    b. absorption spectrum.
    c. emission line.
    d. partial spectrum.

_______ **7.** What colors are shown in a continuous spectrum?
    a. primary colors
    b. cool colors
    c. warm colors
    d. all colors

Directed Reading B *continued*

______ **8.** The colors that appear when a chemical element emits light are called
    a. continuous lines.
    b. absorption lines.
    c. color lines.
    d. emission lines.

______ **9.** Each element in a hot gas can be identified by
    a. a unique set of bright emission lines.
    b. a unique set of bright absorption lines.
    c. a set of emission lines shared with other elements.
    d. a set of absorption lines shared with other elements.

10. Why is the spectrum of a star called an *absorption spectrum*?

____

____

____

11. How is an absorption spectrum produced?

____

____

____

12. What do the black lines of a star's spectrum represent?

____

____

13. In what ways is the pattern of lines in a star's absorption spectrum unique?

____

____

14. Why is it often difficult to identify a star's elements from its absorption spectrum?

____

____

15. What are the two main elements found in stars?

____

____

Directed Reading B *continued*

16. What are the three most common trace elements found in stars?

____

____

____

CLASSIFYING STARS

_______ **17.** In the 1800s, astronomers classified stars according to
   a. their elements.
   b. their temperature.
   c. their age.
   d. their size.

_______ **18.** Stars are now classified by
   a. their elements.
   b. their temperature.
   c. their age.
   d. their size.

_______ **19.** Class O stars, the hottest stars, are
   a. yellow.
   b. orange.
   c. red.
   d. blue.

20. Early astronomers called the brightest stars in the sky

_______________________ stars.

21. What type of numbers are used to represent the magnitudes of dim stars?

____

22. What type of numbers are used to represent the magnitudes of very bright stars?

____

HOW BRIGHT IS THAT STAR?

23. The brightness of a star as seen from Earth is its _______________________.

24. The brightness that a star would have at a distance of 32.6 light-years from

Earth is its _______________________.

Directed Reading B *continued*

25. Why is the sun the brightest object in the sky?

____

____

____

____

DISTANCE TO THE STARS

______ **26.** What unit of measurement do astronomers use to measure the distance from Earth to the stars?
   a. a solar year
   b. a parallax
   c. a light-year
   d. magnitude

27. The distance that light travels in one year; about 90.46 trillion kilometers, is

called a(n) ___________________.

28. An apparent shift in the position of an object when seen from different

locations is called ___________________.

MOTIONS OF STARS

29. Explain why you see different constellations in the sky at different times of the year.

____

____

____

30. What causes the stars to appear to make one complete circle around Polaris every 24 hours?

____

31. Why is the actual motion of the stars difficult to see?

____

____

____

Directed Reading B

Section: The Life Cycle of Stars
TYPES OF STARS

1. List seven ways that stars are classified.

____

____

____

____

2. Why would the classification of a star change as it ages?

____

____

THE LIFE CYCLE OF SUNLIKE STARS

Match the correct description with the correct term. Write the letter in the space provided.

_______ **3.** the first stage of a star's life cycle, when gravity pulls gas and dust into a sphere

_______ **4.** the second and longest stage of a star's life cycle; energy is generated in its core

_______ **5.** a star uses all of its hydrogen, the center of the star shrinks, and the atmosphere grows large and cools

_______ **6.** the final stage of a star's life cycle; the leftover center of a red giant no longer generates energy by nuclear fusion

a. red giant or red supergiant

b. white dwarf

c. main-sequence

d. protostar

A TOOL FOR STUDYING STARS

_______ **7.** The H-R diagram shows the relationship between a star's surface temperature and its
a. absolute magnitude.
b. color.
c. apparent magnitude.
d. age.

Directed Reading B *continued*

8. Where is temperature indicated on the H-R diagram?

____

9. Where is absolute magnitude indicated on the H-R diagram?

____

10. Where does a star spend most of its lifetime as indicated on the H-R diagram?

____

Match the correct description with the correct term. Write the letter in the space provided.

______ **11.** part of the H-R diagram where hot (blue) stars are indicated

______ **12.** part of the H-R diagram where cool (red) stars are indicated

______ **13.** part of the H-R diagram where bright stars are indicated

______ **14.** part of the H-R diagram where dim stars are indicated

a. right side

b. top

c. left side

d. bottom

15. As they age, how do main-sequence stars move on the H-R diagram?

____

____

____

____

Directed Reading B *continued*

THE AGING OF MASSIVE STARS

Match the correct description with the correct term. Write the letter in the space provided.

______ **16.** a gigantic explosion in which a massive star collapses and throws its outer layers into space

______ **17.** the center of a collapsed star contracts into a small, dense ball of neutrons

______ **18.** a spinning neutron star sends out beams of radiation that sweep across space

______ **19.** the contraction of a collapsed star leaves an object so dense and massive that light cannot escape its gravity

a. neutron star

b. black hole

c. supernova

d. pulsar

20. How are black holes detected by astronomers?

____

____

Directed Reading B

Section: Galaxies

1. A collection of stars, dust, and gas bound together by gravity is called

a(n) _________________________________.

TYPES OF GALAXIES

Each of the following statements is true of a spiral galaxy, an elliptical galaxy, or an irregular galaxy. Write *S* for a spiral galaxy, *E* for an elliptical galaxy, and *I* for an irregular galaxy in the space provided.

______ **2.** These galaxies have stopped making stars.

______ **3.** The Milky Way is this type of galaxy.

______ **4.** Some of these are formed when galaxies collide.

______ **5.** The arms of these galaxies are made up of gas, dust, and new stars.

______ **6.** These galaxies are round or oval, like cosmic snowballs.

______ **7.** These galaxies can contain from 10 million to several billion stars.

______ **8.** These galaxies have a bulge in the center and spiral arms.

CONTENTS OF GALAXIES

______ **9.** A large cloud of gas and dust in interstellar space where stars are born is called a(n)
    a. globular cluster.
    b. open cluster.
    c. quasar.
    d. nebula.

______ **10.** A highly concentrated group of up to one million stars is called a(n)
    a. globular cluster.
    b. open cluster.
    c. quasar.
    d. nebula.

______ **11.** A relatively close group of up to 100 to 1,000 stars, usually located along the disk of a spiral galaxy, is called a(n)
    a. globular cluster.
    b. open cluster.
    c. quasar.
    d. nebula.

Directed Reading B *continued*

Quasars

12. Starlike sources of energy located in the centers of galaxies are called

_______________________.

ORIGIN OF GALAXIES

13. Why is looking through a telescope like looking back through time?

____

____

____

14. Why do scientists study distant galaxies?

____

____

____

Skills Worksheet

Directed Reading B

Section: Formation of the Universe

1. The study of the origin, structure, and future of the universe is

called ___________________.

THE BIG BANG THEORY

_______ **2.** What have scientists learned by studying the movement of galaxies?
   a. Most galaxies are moving apart, and the universe is expanding.
   b. Most galaxies are getting closer together, and the universe is getting smaller.
   c. Galaxies do not move, and the universe is not expanding.
   d. Galaxies rotate within the same portion of the universe.

_______ **3.** The standard model used to explain the expansion of the universe is
   a. the theory of universal expansion.
   b. the theory of cosmology.
   c. the theory of fundamental forces.
   d. the big bang theory.

4. The theory that all matter and energy was compressed into an extremely small volume billions of years ago, then exploded and expanded in all

directions, is ___________________.

5. According to the big bang theory, about how long ago did the universe begin?

____

6. Describe *cosmic background radiation.*

____

____

____

____

____

____

____

Directed Reading B *continued*

GRAVITY AND THE UNIVERSE

7. How does gravity act on matter and galaxies to control the size and shape of the universe?

____

____

____

8. What are three structures in the universe to which Earth belongs?

____

____

____

HOW OLD IS THE UNIVERSE?

______ **9.** The oldest stars in the Milky Way galaxy are called
   a. protostars.
   b. blue stars.
   c. yellow stars.
   d. white dwarfs.

10. After studying white dwarf stars, why do scientists believe the universe must be approximately 14 billion years old?

____

____

____

A FOREVER-EXPANDING UNIVERSE

______ **11.** What makes up the smallest amount of total matter in the universe?
   a. dark energy
   b. dark matter
   c. matter making up stars and planets
   d. hydrogen

______ **12.** How can dark matter be detected?
   a. by its light
   b. by its temperature
   c. by its elements
   d. by its gravity

Directed Reading B *continued*

_______ **13.** Most of the universe is composed of
      a. stars and planets.
      b. light energy.
      c. dark energy.
      d. dark matter.

14. What does dark energy seem to be doing?

____

____

____

15. Describe what might happen if the expansion rate of the universe continues to grow.

____

____

____

Skills Worksheet)

Vocabulary and Section Summary A

Stars

VOCABULARY

In your own words, write a definition of the following terms in the space provided.

1. spectrum

____

____

2. apparent magnitude

____

____

3. absolute magnitude

____

____

4. light-year

____

____

5. parallax

____

____

SECTION SUMMARY

Read the following section summary.

- The color of a star depends on the temperature of the star. Blue stars are hottest. Red stars are coolest.

- The spectrum of a star shows which elements make up a star's atmosphere.

- Apparent magnitude is the brightness of a star as seen from Earth. Absolute magnitude is a measure of how bright a star would be if the star were 32.6 light-years from Earth.

- Astronomers use parallax and trigonometry to measure distances to stars that are close to Earth. They use light-years to describe those distances.

- Stars appear to move because of Earth's rotation. The actual motion of stars is hard to see because stars are so distant.

Vocabulary and Section Summary A

The Life Cycle of Stars

VOCABULARY

In your own words, write a definition of the following terms in the space provided.

1. main sequence

____

____

____

2. H-R diagram

____

____

____

3. supernova

____

____

____

SECTION SUMMARY

Read the following section summary.

- New stars form from gas and dust, which are pulled into a sphere by gravity.

- Some types of stars include main-sequence stars, giants, super giants, and white dwarfs.

- Most stars, including the sun, are main-sequence stars.

- The H-R diagram shows the brightness of a star relative to the temperature of the star. It also shows the life cycle of stars.

- Massive stars can explode in a large, bright flash called a *supernova*. Their cores can change into neutron stars or black holes.

Skills Worksheet

Vocabulary and Section Summary A

Galaxies

VOCABULARY

In your own words, write a definition of the following terms in the space provided.

1. galaxy

____

____

____

2. nebula

____

____

____

SECTION SUMMARY

Read the following section summary.

- Astronomers classify galaxies by shape. The three types of galaxies are spiral galaxies, elliptical galaxies, and irregular galaxies.

- Some galaxies contain nebulas and star clusters.

- A nebula is a cloud of gas and dust. A globular cluster is a highly concentrated group of stars. An open cluster is a group of stars that are relatively close together.

- Scientists look at distant galaxies to see what early galaxies looked like.

Vocabulary and Section Summary A

Formation of the Universe

VOCABULARY

In your own words, write a definition of the following term in the space provided.

1. big bang theory

____

____

____

SECTION SUMMARY

Read the following section summary.

- According to the big bang theory, the universe began with a tremendous explosion about 14 billion years ago.
- The presence of cosmic background radiation helps support the big bang theory.
- Scientists use white dwarfs to estimate the age of the universe.
- The universe is composed of matter, dark matter, and dark energy.
- Scientists think that the universe may expand forever.

Skills Worksheet)

Vocabulary and Section Summary B

Stars

VOCABULARY

After you finish reading the section, try this puzzle! In each of the following items, use the clue to unscramble the letters, and write the term in the corresponding blanks.

1. the band of colors produced when white light passes through a prism:
   CRTESMPU

   ___ ___ ___ ___ ___ ___ ___ ___

2. the brightness of a star as seen from Earth: PATAPENR MTGEINAUD

   ___ ___ ___ ___ ___ ___ ___ ___   ___ ___ ___ ___ ___ ___ ___ ___ ___

3. the brightness that a star would have at a distance of 32.6 light-years from Earth: LUBSETOA GMIEDUATN

   ___ ___ ___ ___ ___ ___ ___ ___   ___ ___ ___ ___ ___ ___ ___ ___ ___

4. the distance that light travels in one year; about 9.46 trillion kilometers:
   ILTGH-AREY

   ___ ___ ___ ___ ___ - ___ ___ ___ ___

5. an apparent shift in the position of an object when viewed from different locations: AARXPLLA

   ___ ___ ___ ___ ___ ___ ___ ___

SECTION SUMMARY

Read the following section summary.

- The color of a star depends on the temperature of the star. Blue stars are hottest. Red stars are coolest.

- The spectrum of a star shows which elements make up a star's atmosphere.

- Apparent magnitude is the brightness of a star as seen from Earth. Absolute magnitude is a measure of how bright a star would be if the star were 32.6 light-years from Earth.

- Astronomers use parallax and trigonometry to measure distances to stars that are close to Earth. They use light-years to describe those distances.

- Stars appear to move because of Earth's rotation. The actual motion of stars is hard to see because stars are so distant.

Vocabulary and Section Summary B

The Life Cycle of Stars
VOCABULARY

After you finish reading the section, try this puzzle! In the space provided, write the term described. Then, find the words in the word search puzzle on the next page. Terms can be hidden vertically, horizontally, diagonally, or backward.

_______________________ **1.** the location on the H-R diagram where most stars lie; has a diagonal pattern from the lower right (low temperatures and luminosity) to the upper left (high temperature and luminosity)

_______________________ **2.** Hertzsprung-Russell diagram; a graph that shows the relationship between a star's surface temperature and absolute magnitude

_______________________ **3.** a gigantic explosion in which a massive star collapses and throws its outer layers into space

_______________________ **4.** following a supernova, the center of the collapsed star that has contracted into a very small but very dense ball of neutrons

_______________________ **5.** a spinning neutron star

_______________________ **6.** a star that has collapsed into an object so dense and massive that light cannot escape its gravity

Vocabulary and Section Summary B *continued*

T	M	W	H	U	B	E	L	Q	Y	N	I	T	O	T
P	H	B	G	P	C	N	K	O	F	E	K	J	Z	Z
E	B	K	Q	W	N	N	V	V	D	U	M	Y	Q	L
N	E	H	V	S	P	B	R	X	B	T	H	F	Q	Y
T	Y	I	M	R	U	Q	Z	O	I	R	F	D	W	O
M	N	B	A	B	L	A	C	K	H	O	L	E	F	M
K	P	H	D	U	S	U	P	E	R	N	O	V	A	I
M	M	G	U	Y	A	S	Q	J	I	S	O	R	J	W
W	X	S	P	S	R	N	F	B	N	T	G	Q	H	C
E	C	N	E	U	Q	E	S	N	I	A	M	Q	Q	M
L	D	P	J	N	Y	P	M	D	I	R	M	B	C	U
J	A	L	A	C	W	G	Z	D	E	T	J	L	N	X
U	V	W	T	T	P	U	R	M	H	A	Q	U	X	X
A	W	M	W	C	T	H	P	R	U	T	T	V	M	H
K	O	S	V	P	W	J	F	J	Z	X	E	D	A	C

SECTION SUMMARY

Read the following section summary.

- New stars form from gas and dust, which are pulled into a sphere by gravity.
- Some types of stars include main-sequence stars, giants, super giants, and white dwarfs.
- Most stars, including the sun, are main-sequence stars.
- The H-R diagram shows the brightness of a star relative to the temperature of the star. It also shows the life cycle of stars.
- Massive stars can explode in a large, bright flash called a *supernova*. Their cores can change into neutron stars or black holes.

Skills Worksheet

Vocabulary and Section Summary B

Galaxies
VOCABULARY

After you finish reading the section, try this puzzle! Use the clues below to solve the crossword puzzle below.

ACROSS

5. a galaxy with a bulge at its center and spiral arms that are made up of gas, dust, and new stars

7. a large cloud of gas and dust in interstellar space; a region in space where stars are born

DOWN

1. groups of 100 to 1,000 stars that are close together relative to other stars

2. a highly concentrated group of stars that looks like a ball

3. a pair of irregular galaxies in the neighborhood of the Milky Way

4. the spiral galaxy in which the sun is located

6. a collection of stars, dust, and gas bound together by gravity

Vocabulary and Section Summary B *continued*

SECTION SUMMARY

Read the following section summary.

- Astronomers classify galaxies by shape. The three types of galaxies are spiral galaxies, elliptical galaxies, and irregular galaxies.

- Some galaxies contain nebulas and star clusters.

- A nebula is a cloud of gas and dust. A globular cluster is a highly concentrated group of stars. An open cluster is a group of stars that are relatively close together.

- Scientists look at distant galaxies to see what early galaxies looked like.

Skills Worksheet

Vocabulary and Section Summary B

Formation of the Universe
VOCABULARY

After you finish reading the section, try this puzzle! Look at the clues below, and write the terms being described in the blanks on the next page. Then, arrange the letters in the boxes to answer the question.

1. very old stars that are studied to estimate the age of the universe

2. a component of the universe that seems to be counteracting the effect of gravity and accelerating the expansion of the universe

3. the study of the origin, structure, processes, and evolution of the universe

4. radiation left over from the big bang and detectable as radio "noise" coming from all directions in space

5. a component of the universe that does not give off light but has gravity

6. the theory that all matter and energy in the universe was compressed into an extremely small volume that, 13 billion to 15 billion years ago, exploded and began expanding in all directions

1. _ _ _ ☐ _ _ _ _ _ _ ☐

2. _ ☐ _ _ _ _ _ ☐ _ _

3. ☐ _ _ _ _ ☐ _ _ _

4. _ _ ☐ _ _ _ _ _ _ _ _ _ _ ☐ _ _

   _ _ _ _ _ _ _ _

5. _ _ _ _ _ _ ☐ _ _ _

6. _ _ _ _ _ _ _ _ _ ☐ _ ☐ _

7. What is the term?

____

Vocabulary and Section Summary B *continued*

SECTION SUMMARY

Read the following section summary.

- According to the big bang theory, the universe began with a tremendous explosion about 14 billion years ago.
- The presence of cosmic background radiation helps support the big bang theory.
- Scientists use white dwarfs to estimate the age of the universe.
- The universe is composed of matter, dark matter, and dark energy.
- Scientists think that the universe may expand forever.

Reinforcement

Diagramming the Stars

Complete this worksheet after you finish reading the section "The Life Cycle of Stars."

An H-R diagram shows the relationship between a star's surface temperature and its absolute magnitude. Follow the instructions below to create your own H-R diagram on the next page. You may want to use colored pencils or crayons for this activity. Remember that a star's brightness increases as you move toward the top of the H-R diagram.

1. Our sun is an average star. It should be located at about the center of the diagram. Draw and label the sun on the diagram.

2. Draw and label a red dwarf star on the diagram. Red dwarf stars are dim and have a low temperature.

3. Draw and label a white dwarf star on your diagram. White dwarf stars are dim and have a high temperature.

4. Draw and label a blue star on the diagram. Blue stars are very hot and bright.

5. Draw and label a red giant on the diagram. Red giants are cool and bright.

6. Most stars can be plotted along the main sequence of an H-R diagram. These stars range from very bright, very hot stars to dim, cool stars. Indicate and label on your diagram where the main sequence should go.

7. Which of the stars that you have plotted are included in the main sequence?

____

____

____

8. Imagine that you have discovered a new star in the night sky. Your measurements show that it has a surface temperature of 10,000°C and an absolute magnitude of +10. Based on your diagram, what type of star do you think it is?

____

____

____

Reinforcement *continued*

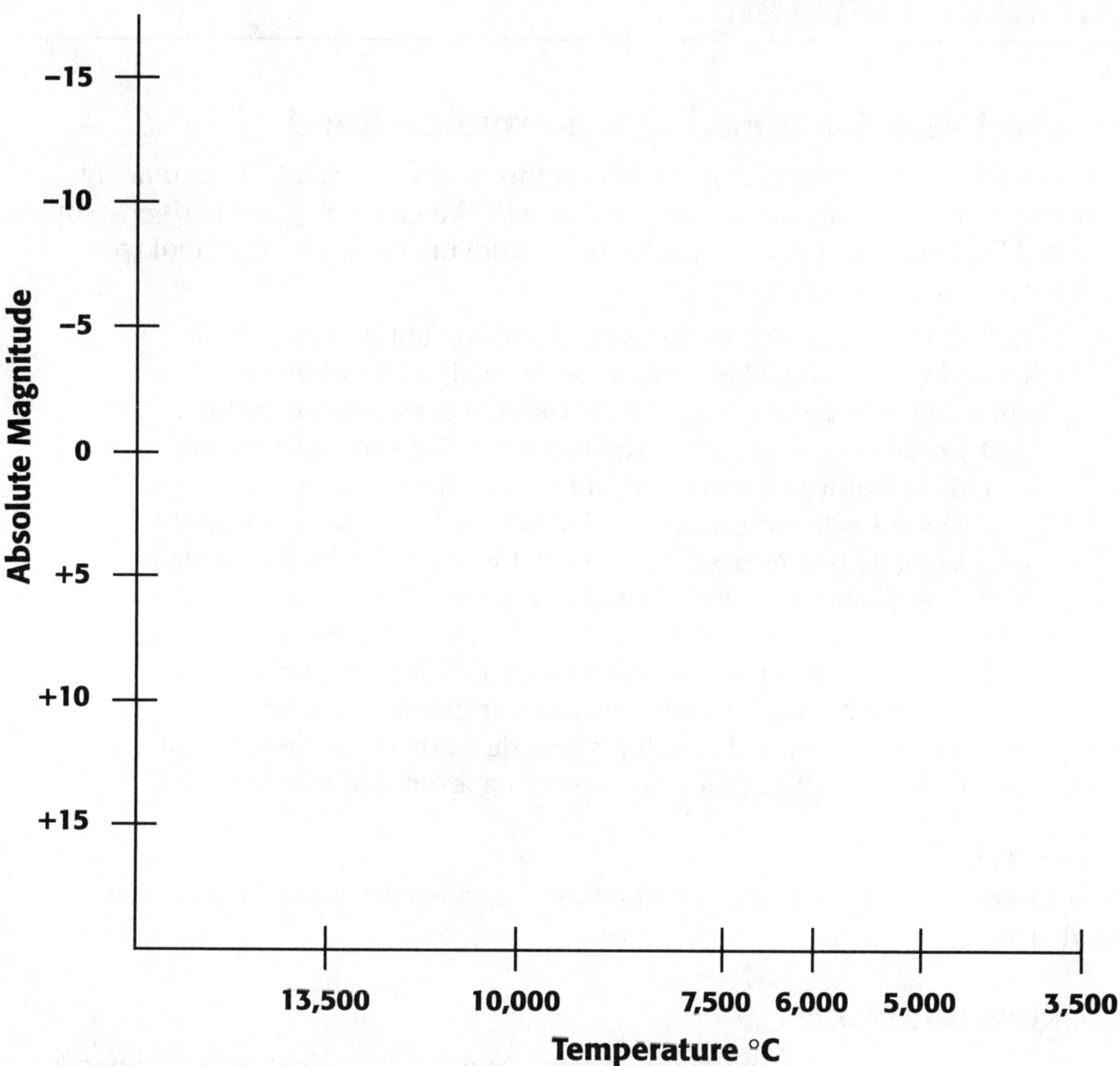

Name _________________________________ Class _______________ Date _____________

Critical Thinking

Fleabert and the Amazing Watermelon Seed

One day you spot a dirty, tattered book on the side of the road. Overcome by curiosity, you pick it up and realize the book is *The Origin of the Universe*, by Psoodry Psantalando. Because you've been studying the topic at school, you decide to read on. Here's an excerpt:

> Before the beginning, there existed nothing but a seed (which looked remarkably like a watermelon seed) and a clumsy spirit-hamster named Fleabert. Within the seed was everything that would ever be a part of the universe. The seed was packed so tightly that it was too heavy for even a spirit-hamster to move.
>
> Fleabert was curious about this little seed because it was the only thing he had ever seen. One day Fleabert carelessly nibbled at the seed, and it exploded right in his face. The contents of the seed burst forth and began to spread in every direction. The seed began to grow into the great shrubbery that is now our universe.
>
> This shrubbery, with trillions upon trillions of leaves, continues to grow and develop every day. Some people even say that they can still see Fleabert scurrying around in the heavens.

USEFUL TERMS
shrubbery short, woody plants with several permanent stems instead of a single trunk

SEEING RELATIONSHIPS

1. If the universe can be understood as shrubbery, what might the shrubbery's branches represent?

____

____

2. What could the shrubbery's leaves represent?

____

____

Critical Thinking *continued*

ANALYZING A VIEWPOINT

3. Would Psoodry Psantalando's story of the formation of the universe be considered scientific? Explain your answer.

____

____

____

MAKING COMPARISONS

4. Does this explanation of the origin of the universe have anything in common with the big bang theory? Explain your answer.

____

____

____

____

5. What major question about the beginning of the universe does this explanation answer that the big bang theory fails to answer?

____

____

____

____

DRAWING CONCLUSIONS

6. Do you think that science will eventually provide the answers to all of our questions about the origin of the universe? Explain your answer.

____

____

____

____

SciLinks Activity

GALAXIES

Go to www.scilinks.org. To find links related to galaxies, type in the keyword HY71448. Then, use the links to complete the following activity about galaxies.

Internet Resources

For a variety of links related to this chapter, go to www.scilinks.org

Topic: Stars
SciLinks code: HY71448

Imagine that you are creating a book of space photographs. You have pictures of supernovas, red giants, white dwarfs, pulsars, and black holes, but none of galaxies. Browse through the sites on galaxies and view the many images available. Select three images—one for each of the three major types of galaxies. Describe each galaxy in the spaces below by drawing a picture or writing a paragraph. Label each description with the name of the galaxy.

Tip: Look for defining characteristics, such as the spiral arms of spiral galaxies, and be sure to include them in your descriptions.

Section Review

Stars

USING VOCABULARY

1. Use *apparent magnitude* and *absolute magnitude* in the same sentence.

____

____

2. Use *spectrum*, *light-year*, and *parallax* in separate sentences.

____

____

UNDERSTANDING CONCEPTS

3. Analyzing Explain how color indicates the temperature of a star.

____

____

____

4. Summarizing How do astronomers use spectra to determine the composition of a star?

____

____

____

5. Identifying What two elements are most commonly found in stars?

____

____

____

6. Comparing Compare the absolute magnitude of stars with the apparent magnitude of stars.

____

____

Section Review *continued*

7. Summarizing How do astronomers measure distances from Earth to stars?

____

____

____

8. Analyzing Explain why astronomers use light-years to describe distances from Earth to stars.

____

____

____

9. Summarizing Explain why stars appear to move in the sky.

____

____

____

CRITICAL THINKING

10. Making Comparisons Compare a continuous spectrum with an absorption spectrum. Then, explain how an absorption spectrum can identify the elements in a star's atmosphere.

____

____

____

____

11. Identifying Relationships The apparent magnitude of the sun is −26.8, while the absolute magnitude of the sun is +4.8. Explain why.

____

____

____

Section Review *continued*

12. Applying Concepts If a certain star displayed a large parallax, what could you say about the star's distance from Earth?

____

____

____

MATH SKILLS

13. Making Calculations How many times as bright as a light that is 90 m away from you does a light that is 10 m away from you appear? Show your work below.

CHALLENGE
INTERPRETING GRAPHICS

Use the illustrations below to answer the next question.

14. Evaluating Data If all of the stars appear to make one complete circle around Polaris every 24 h, approximately how much time has passed between the illustration of the sky on the left and the illustration of the sky on the right?

____

____

____

____

Section Review

The Life Cycle of Stars

USING VOCABULARY

1. Use *H-R diagram* and *main sequence* in the same sentence.

____

____

UNDERSTANDING CONCEPTS

2. Applying Explain the process by which new stars form.

____

____

____

3. Arranging Arrange the following stars in the order that reflects the life cycle of a star: white dwarf, red giant, and main sequence star.

____

____

____

4. Analyzing In main-sequence stars, how does brightness relate to temperature?

____

____

____

5. Applying Explain what happens when a massive star ends its life as a supernova.

____

____

____

6. Comparing Compare a neutron star and a pulsar.

____

____

____

| Section Review *continued*

7. Summarizing How can the center of a collapsed star form a black hole?

____

____

CRITICAL THINKING

8. Evaluating Data How does the H-R diagram explain the life cycle of a star?

____

____

9. Identifying Relationships Why do massive stars have shorter lives than other stars do?

____

____

10. Analyzing Processes Describe what might happen to a star after it explodes as a supernova.

____

____

____

CHALLENGE

11. Evaluating Hypotheses An astronomer hypothesizes that gas and dust from a star are being drawn into a black hole. How can the astronomer test his or her hypothesis to see if the hypothesis is correct?

____

____

____

____

Section Review

Galaxies

USING VOCABULARY

1. Use the following terms in the same sentence: *galaxy* and *nebula*.

____

____

UNDERSTANDING CONCEPTS

2. Summarizing Briefly describe the differences between spiral, elliptical, and irregular galaxies.

____

____

____

____

3. Listing List four facts about the Milky Way. Include the location of the sun in the Milky Way.

____

____

____

____

4. Comparing Compare a nebula with a star cluster.

____

____

____

5. Identifying In which types of galaxy are globular clusters most likely to be found?

____

____

____

Section Review *continued*

6. Applying Explain how scientists see what early galaxies looked like.

____

____

____

____

CRITICAL THINKING

7. Making Comparisons Describe the difference between a globular cluster and an open cluster.

____

____

____

____

8. Identifying Relationships Explain how looking through a telescope is like looking back in time.

____

____

____

____

CHALLENGE

9. Applying Concepts Explain why looking at distant galaxies gives scientists an idea about how galaxies change over time.

____

____

____

____

____

____

Section Review

Formation of the Universe

USING VOCABULARY

1. Write an original definition for the *big bang theory*.

____

____

UNDERSTANDING CONCEPTS

2. Summarizing How is gravity related to the big bang?

____

____

____

____

3. Applying How does the presence of cosmic background radiation support the big bang theory?

____

____

____

____

4. Summarizing Describe the structure of the universe.

____

____

____

5. Analyzing What will happen to the universe if it expands forever?

____

____

____

CRITICAL THINKING

6. Analyzing Ideas Explain why astronomers use white dwarfs to date the universe.

____

____

____

____

____

____

INTERPRETING GRAPHICS

Use the illustration below to answer the next question.

7. Evaluating Data If distances AB, AC, and AD double, how fast will raisins B, C, and D be moving away from raisin A? Answer in centimeters per second (cm/s).

____

____

____

____

____

Skills Worksheet

Chapter Review

USING VOCABULARY

_______ **1. Academic Vocabulary** Which of the following words is the closest in meaning to the word *structure*?
    a. outline   **c.** arrangement
    b. surface   **d.** shape

Correct each statement by replacing the underlined term.

2. <u>Absolute magnitude</u> is the brightness of a star as seen from Earth.

____

3. The distance that light travels in space in 1 year is called <u>parallax</u>.

____

4. The <u>main sequence</u> is a graph that shows the relationship between the surface temperature and absolute magnitude of a star.

____

UNDERSTANDING CONCEPTS
Multiple Choice

_______ **5.** Which of the following stars has the highest surface temperature?
    a. a red star   **c.** a yellow star
    b. a blue star   **d.** an orange star

_______ **6.** Approximately how many kilometers does a light-year equal?
    a. 9.46 trillion   **c.** 9.46 quadrillion
    b. 9.46 million   **d.** 9.46 billion

_______ **7.** The sun is just one of many stars in the Milky Way. Which type of star is the sun?
    a. a neutron star   **c.** a main-sequence star
    b. a white dwarf   **d.** a red giant

_______ **8.** The Milky Way is an example of a(n)
    a. elliptical galaxy.   **c.** irregular galaxy.
    b. lenticular galaxy.   **d.** spiral galaxy.

_______ **9.** Which of the following contain billions of stars and may have different shapes?
    a. open clusters   **c.** nebulas
    b. galaxies   **d.** globular clusters

| Chapter Review *continued*

Short Answer

10. Classifying Describe how scientists classify stars.

____

____

11. Applying Explain why astronomers use light-years to estimate the distance between Earth and stars.

____

____

12. Listing List the stages in the life cycle of a star in the order in which they happen.

____

____

____

13. Analyzing Explain the role of nuclear fusion in the life cycle of a star.

____

____

14. Comparing Compare spiral galaxies, elliptical galaxies, and irregular galaxies.

____

____

____

15. Describing Describe the Milky Way, including the sun's position in the Milky Way.

____

____

____

16. Applying Explain how the presence of cosmic background radiation supports the big bang theory.

____

____

____

| Chapter Review *continued*

17. Comparing Compare dark matter with dark energy.

____

____

WRITING SKILLS

18. Creative Writing Imagine that you are an astronomer who has made a new discovery either in the Milky Way or in another part of the universe. You may have discovered a star, a star cluster, a nebula, a black hole, or a galaxy. Describe your discovery on a separate piece of paper in the form of a short story. Be sure to describe the way in which you made your discovery.

CRITICAL THINKING

19. Concept Mapping Use the following terms to create a concept map: *main sequence star, nebula, red giant, white dwarf, neutron star,* and *black hole.*

20. Evaluating Data You are looking at a star through a telescope in January. There is a small shift in the star's apparent position compared to when you looked at the same star through a telescope in July. What can you conclude about the distance of the star from Earth?

____

____

____

____

____

21. Making Comparisons Explain the differences between main-sequence stars, giant stars, and white dwarfs.

____

____

____

____

22. Evaluating Conclusions While looking through a telescope, you see a galaxy that doesn't appear to contain any blue stars. What kind of galaxy is the galaxy most likely to be? Explain your answer.

____

____

____

____

____

23. Evaluating Sources According to the big bang theory, how did the universe begin?

____

____

____

____

INTERPRETING GRAPHICS

Use the graph below to answer the next two questions.

Relationship Between Age and Mass of a Star

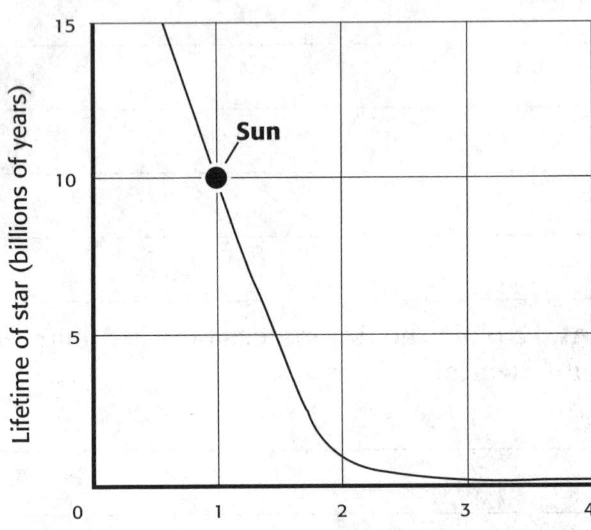

Mass of star (compared to sun)

24. **Identifying Relationships** Which star would live longer: a star that has half the mass of the sun or a star that has 2 times the mass of the sun? Explain your answer.

____

____

____

____

____

25. **Applying Concepts** Approximately how long would a main-sequence star that has a mass of about 1.5 times the mass of the sun live?

____

____

____

____

____

INTERPRETING GRAPHICS

The graph below shows Hubble's law, which relates how far galaxies are from Earth and how fast they are moving away from Earth.

Use the graph below to answer the next two questions.

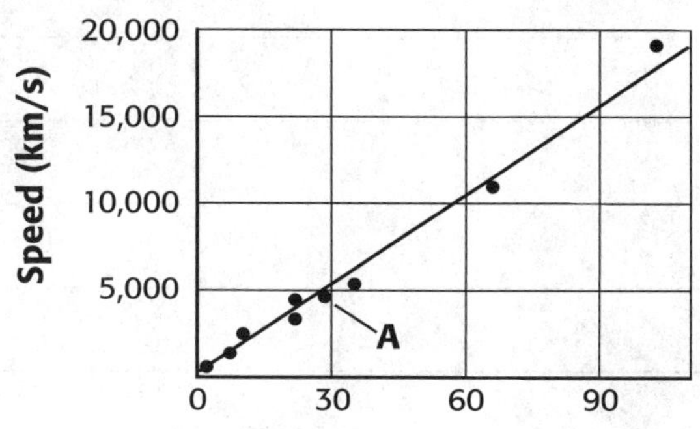

26. **Evaluating Data** If a galaxy is moving away from Earth at 15,000 km/s, how far from Earth is the galaxy?

____

____

____

27. **Evaluating Data** If a galaxy is 60 million light-years from Earth, how fast is it moving away from Earth?

____

____

____

MATH SKILLS

28. **Making Calculations** One star has a magnitude of 25, and another star has a magnitude of 115. How much brighter than the star that has the magnitude of 115 is the star that has the magnitude of 25? Show your work below.

CHALLENGE

29. Applying Concepts There are more low-mass stars than high-mass stars in the universe. Given this information, do you think that more stars will end their lives as white dwarfs or in supernovas?

____

____

____

____

____

____

____

____

30. Making Inferences What is the relationship between gravity and the unknown material that astronomers call *dark energy*?

____

____

____

____

____

____

____

____

Chapter Pretest

Teacher Notes and Answer Key

The Pretest questions are designed to help you determine the prior knowledge of your students. Some questions test whether students have mastered the background knowledge they need to understand the content you are about to teach. Other questions test your students' prior knowledge of the content you are about to teach. Use the Pretest with the Test Doctors and diagnostic teaching tips in these notes pages to help you tailor your instruction to your students' specific needs.

QUESTION NUMBER	CORRECT ANSWER	STANDARD
1	A	6.4.b
2	D	7.6.a
3	C	7.6.e
4	B	8.2.d
5	A	8.7.b
6	B	8.2.g
7	C	8.4.a
8	D	8.4.b
9	A	8.4.b
10	A	8.4.c

TEST DOCTOR

The following Pretest questions have been diagnosed by the Test Doctor. Find out what might be causing your students' "ailing" answers. Each Test Doctor is followed by a diagnostic teaching tip to help you address students' learning needs.

Question 1 *asks students to identify a fact about the sun's energy.*

A Correct. The sun's energy reaches Earth in the form of radiation.

B Incorrect. The sun's energy is not received on Earth as sound.

C Incorrect. The sun's energy reaches all planets in our solar system, not just Earth.

D Incorrect. Much of the sun's energy is absorbed by Earth's atmosphere.

Diagnostic Teaching Tip: Students who have difficulty with this question should review how the atmosphere affects the amount of radiation Earth receives. Have students work in pairs to research solar radiation. Have partners make a pie chart showing what happens to the radiation after it reaches Earth's atmosphere. (Approximately 25% is scattered and reflected by clouds and air; 20% is absorbed by ozone, clouds, and gases; 50% is absorbed by Earth's surface; and 5% is reflected by Earth's surface.)

Question 2 *asks students to identify the narrowest wavelength range in the electromagnetic spectrum.*

A Incorrect. The gamma ray part of the spectrum has a much larger wavelength range than does visible light.

Chapter Pretest *continued*

B Incorrect. Infrared has a much larger wavelength range than does visible light.

C Incorrect. Ultraviolet has a larger wavelength range than does visible light.

D Correct. Visible light has the narrowest wavelength range in the electromagnetic spectrum.

Diagnostic Teaching Tip: Students who have difficulty answering this question should review the electromagnetic spectrum. Remind students that radiation travels through space in the form of waves at a very high speed. The electromagnetic spectrum contains all of the types of electromagnetic waves. Have students focus on the visible light area of the spectrum.

Question 3 *asks students to understand why white light is visible.*

A Incorrect. There is no wavelength for black light.

B Incorrect. The human eye sees wavelengths for all colors in the visible spectrum.

C Correct. The human eye sees all visible colors in the spectrum, which together make up white light.

D Incorrect. The human eye sees all visible colors in the spectrum, which together make up white light.

Diagnostic Teaching Tip: Students who have difficulty with this question should review the visible light portion of the electromagnetic spectrum. Have students study the different wavelengths (colors) that make up white light.

Question 4 *asks students to know how different forces act on an object.*

A Incorrect. Air resistance will slow the ball down but it will not make it drop.

B Correct. The force of gravity causes a ball to fall regardless of its speed or direction.

C Incorrect. The speed of a ball thrown horizontally will not make it drop.

D Incorrect. The horizontal curve of the ball's path will not make it drop.

Diagnostic Teaching Tip: Students who have trouble answering this question correctly should review the forces that act on objects. Have students discuss each answer. Discussion should focus on how each will affect a horizontally-thrown ball.

Question 5 *asks students to compare helium and hydrogen.*

A Correct. Helium has two protons and hydrogen has one proton.

B Incorrect. Helium has two neutrons and hydrogen has no neutrons.

C Incorrect. Helium has two electrons and hydrogen has one electron.

D Incorrect. Both helium and hydrogen have two stable isotopes.

Diagnostic Teaching Tip: Students who have difficulty answering this question correctly might benefit from a review of common elements of the periodic table. Call out a common element on the periodic table. Ask volunteers to give the number of protons, number of neutrons, and number of electrons for the element. Be sure to cover helium and hydrogen in the review.

| Chapter Pretest *continued*

Question 6 *asks students to realize that gravity is responsible for forming galaxies after the big bang.*

A **Incorrect.** Fusion occurs within a star, but it does not form galaxies.

B **Correct.** Scientists believe gravity caused the universe to form galaxies after the big bang.

C **Incorrect.** Heat energy is produced by a star, but it does not form galaxies.

D **Incorrect.** Molecular bonds do not form galaxies.

Diagnostic Teaching Tip: Students who answer this question incorrectly might benefit from a simulation. Tell students that gravity is everywhere in the universe, and that it has a great influence on stellar bodies. Scatter iron shavings or iron nails on a piece of paper. Show students a powerful magnet and tell them it represents gravity. Circle the magnet beneath the paper and show how the iron is collected into a circle centered around the magnet. Tell student that this demonstration mimics how gravity caused the matter distributed throughout the universe to form galaxies. Be sure to emphasize Section 4, "Formation of the Universe," in Chapter 15, "Stars, Galaxies, and the Universe." Students must know the role of gravity in forming and maintaining the shapes of galaxies in order to master standard 8.2.g.

Question 7 *asks students to identify the three types of galaxies.*

A **Incorrect.** An elliptical galaxy is a type of galaxy.

B **Incorrect.** An irregular galaxy is a type of galaxy.

C **Correct.** There is no such type of galaxy as a regular galaxy.

D **Incorrect.** A spiral galaxy is a type of galaxy.

Diagnostic Teaching Tip: Students who have difficulty answering this question correctly should benefit from an introduction to galaxies. Show students a photo of each type of galaxy. Ask a student volunteer to tell which type of galaxy Earth's galaxy, the Milky Way, is (spiral). Be sure to emphasize Section 3, "Galaxies," in Chapter 15, "Stars, Galaxies, and the Universe." Students must know that galaxies may have different shapes in order to master standard 8.4.a.

Question 8 *asks students to tell which instrument is used to analyze light given off by a star.*

A **Incorrect.** This is a tool used for studying things that are very small.

B **Incorrect.** A refracting telescope uses lenses that pass light in order to study distant objects, but it does not analyze light.

C **Incorrect.** A reflecting telescope, such as the Hubble Space Telescope, uses mirrors to reflect light in order to study space, but it does not analyze light.

D **Correct.** A spectroscope is used to analyze light given off by a star.

Chapter Pretest *continued*

Diagnostic Teaching Tip: Students who have difficulty answering this question correctly might benefit from a visual introduction to spectra. Show students emission lines given off by elements and tell them that these are unique to an element. If possible, show students the emission lines given off by elements commonly found in stars, such as hydrogen, helium, and calcium. Tell students that scientists study these spectra (singular: spectrum) using an instrument called a spectroscope. Be sure to emphasize Section 1, "Stars," in Chapter 15, "Stars, Galaxies, and the Universe." Students must know that stars may differ in size, color, and temperature in order to master standard 8.4.b.

Question 9 *asks students to identify which color star has the coolest temperature.*

A Correct. Red stars have the coolest temperatures.
B Incorrect. White stars are hot, old stars.
C Incorrect. Blue stars have the hottest temperatures.
D Incorrect. Yellow stars have relatively cool temperatures, but they are not the coolest stars.

Diagnostic Teaching Tip: Students having difficulty with this question should be introduced to the Hertzsprung-Russell (H-R) diagram. Have students look at an H-R diagram or draw one for the class. Point out the groupings of stars on the diagram, given their color and temperature. Be sure to emphasize Section 2, "The Life Cycle of Stars," in Chapter 15, "Stars, Galaxies, and the Universe." Students must know that stars may differ in size, color, and temperature in order to master standard 8.4.b.

Question 10 *asks students to tell which unit is used to measure the distance to stars.*

A Correct. Light-years are used to measure the distance to stars.
B Incorrect. Meters/second are used to measure speed.
C Incorrect. Newtons are used to measure force.
D Incorrect. Kilometers per hour are used to measure speed.

Diagnostic Teaching Tip: Students who have difficulty with this question should view a brief activity using the distance and units of measurement. Tell students that you are going to estimate the width of the classroom using millimeters. Ask students if they think millimeters was a good choice of units to use. Have a student volunteer suggest a more appropriate unit (meter). Tell students that scientists realized that stars were so far away that units used on Earth to measure distance were too small and gave very large numbers. Tell them this is why scientists measure astronomical distance using the distance light travels in a year (light-years) or the distance from the sun to Earth (astronomical units) to measure distances in space. Be sure to emphasize Section 1, "Stars," in Chapter 15, "Stars, Galaxies, and the Universe." Students must know how to use light-years as measures of distance between stars and Earth in order to master standard 8.4.c.

Chapter Pretest

_______ **1.** Which of the following is true about the sun's energy?
  A The sun's energy reaches Earth as radiation.
  B The sun's energy is received on Earth as heat, light, and sound.
  C The sun's energy only reaches Earth, which is why there is no life on other planets.
  D All of the sun's energy reaches Earth's surface from space.

_______ **2.** Which of the following has the narrowest wavelength range in the electromagnetic spectrum?
  A gamma ray
  B infrared
  C ultraviolet
  D visible light

_______ **3.** Why do we see white light?
  A The human eye absorbs the wavelength for black light, allowing humans to see it as white.
  B The human eye sees only the wavelength for white light and does not see wavelengths for other colors.
  C The human eye sees all visible colors in the spectrum, which together make up white light.
  D The human eye sees none of the visible colors in the spectrum, which leaves only white light.

_______ **4.** A ball is thrown horizontally from a Little League pitcher's hand. Which of the following forces causes it to drop downward?
  A air resistance
  B gravity
  C the speed of the ball
  D the horizontal curve of the ball's path

_______ **5.** How do helium and hydrogen, elements that make up blue-white stars such as Rigel, compare?
  A Helium has two protons while hydrogen has only one.
  B Helium has no neutrons but hydrogen has two.
  C Helium has one electron while hydrogen has none.
  D Both helium and hydrogen have no stable isotopes.

_______ **6.** Which of the following do scientists believe caused the universe to form galaxies after the big bang?
  A fusion
  B gravity
  C heat energy
  D molecular bonds

_______ **7.** Which of the following is NOT a type of galaxy?
    A elliptical galaxy
    B irregular galaxy
    C regular galaxy
    D spiral galaxy

_______ **8.** Which of the following is used to analyze the light given off by a star?
    A microscope
    B refracting telescope
    C reflecting telescope
    D spectroscope

_______ **9.** The stars with the coolest temperatures are
    A red.
    B white.
    C blue.
    D yellow.

_______ **10.** Which of the following units do scientists use to measure the distance to stars?
    A light-years
    B meters/second
    C Newtons
    D km/h

Section Quiz

Section: Stars

Write the letter of the correct answer in the space provided.

_______ **1.** The color of a star depends on its
- **a.** size.
- **b.** temperature.
- **c.** shape.
- **d.** magnitude.

_______ **2.** What can a scientist learn about a star from its spectrum?
- **a.** its color
- **b.** its size
- **c.** its composition and temperature
- **d.** its age

_______ **3.** Why do all the stars appear to make one complete circle around Polaris every 24 hours?
- **a.** because of Earth's fixed position
- **b.** because of Earth's rotation
- **c.** because Polaris rotates around Earth
- **d.** because the stars rotate around Earth

_______ **4.** What color are the hottest stars?
- **a.** red
- **b.** yellow
- **c.** orange
- **d.** blue

_______ **5.** Why is the actual motion of stars so difficult to see?
- **a.** Earth rotates counterclockwise.
- **b.** The light of the stars is too bright.
- **c.** The stars are too close to Earth.
- **d.** The stars are very far from Earth.

Match the correct description with the correct term. Write the letter in the space provided.

_______ **6.** the brightness of a star as it would be at a distance of 32.6 light-years from Earth

_______ **7.** the brightness of a star as seen from Earth

_______ **8.** an apparent shift in the position of an object when viewed from different locations

_______ **9.** the distance that light travels in one year

_______ **10.** the band of colors produced when white light passes through a prism

- **a.** light-year
- **b.** spectrum
- **c.** parallax
- **d.** apparent magnitude
- **e.** absolute magnitude

Name _________________________________ Class _______________ Date _____________

Section Quiz

Section: The Life Cycle of Stars

Write the letter of the correct answer in the space provided.

______ **1.** The H-R diagram shows the relationship of a star's surface
temperature and its
- **a.** color.
- **b.** size.
- **c.** apparent magnitude.
- **d.** absolute magnitude.

______ **2.** At about 10,000°C, hydrogen begins to change into helium and energy
is released in a process called
- **a.** a supernova.
- **b.** a pulsar.
- **c.** nuclear fusion.
- **d.** nuclear fission.

______ **3.** A small, hot star near the end of its life is a type of star called a(n)
- **a.** white dwarf.
- **b.** red dwarf.
- **c.** supergiant.
- **d.** protostar.

______ **4.** As gravity pulls gas and dust into a sphere that becomes denser and
hotter, and a sunlike star begins to form, it is called a(n)
- **a.** white dwarf.
- **b.** red dwarf.
- **c.** supergiant.
- **d.** protostar.

______ **5.** What type of star would be shown at the lower right end of the main
sequence on the H-R diagram?
- **a.** white dwarf
- **b.** red dwarf
- **c.** supergiant
- **d.** protostar

**Match the correct description with the correct term. Write the letter in the space
provided.**

______ **6.** a gigantic explosion that causes the death of
a large star

______ **7.** a star in which the electrons and protons
have become neutrons

______ **8.** a star that expands and cools once it runs
out of hydrogen

______ **9.** a spinning neutron star that emits rapid
pulses of radio and optical energy

______ **10.** a massive and dense object from which even
light cannot escape its gravity

- **a.** pulsar
- **b.** neutron star
- **c.** black hole
- **d.** supernova
- **e.** red giant

(Assessment)

Section Quiz

Section: Galaxies

Write the letter of the correct answer in the space provided.

_______ **1.** All of the following are major types of galaxies EXCEPT a(n)

    a. spiral galaxy.   **c.** elliptical galaxy.

    b. irregular galaxy.   **d.** triangular galaxy.

_______ **2.** Where are globular clusters found?

    a. along the spiral disk of galaxies

    b. in the spherical halo of spiral galaxies and near elliptical galaxies

    c. in the center bulge of spiral galaxies

    d. in the spiral arms of spiral galaxies

_______ **3.** Open clusters are found

    a. along the disks of spiral galaxies.

    b. in the spherical halo surrounding spiral galaxies.

    c. in the center bulge of spiral galaxies.

    d. in the cores of some galaxies.

_______ **4.** The galaxy in which Earth is located is the

    a. Andromeda galaxy.   **c.** Milky Way galaxy.

    b. M87 galaxy.   **d.** Magellanic Cloud.

_______ **5.** Why do scientists study distant galaxies?

    a. to learn what galaxies are made of

    b. to learn what early galaxies looked like

    c. to learn about space travel

    d. to learn about the speed of light

Match the correct definition with the correct term. Write the letter in the space provided.

_______ **6.** a collection of billions of stars, dust, and gas bound together by gravity

_______ **7.** a group of 100 to 1,000 closely grouped stars

_______ **8.** a concentrated group of up to a million stars that looks like a ball

_______ **9.** a starlike source of energy located in the center of a galaxy

_______ **10.** a large cloud of gas and dust in interstellar space; a region in space where stars are born

a. nebula

b. galaxy

c. quasar

d. open cluster

e. globular cluster

Assessment

Section Quiz

Section: Formation of the Universe

Write the letter of the correct answer in the space provided.

______ **1.** What have careful measurements shown scientists about the movement of galaxies?
   a. Galaxies are moving apart, and the universe is expanding.
   b. Galaxies are coming together, and the universe is contracting.
   c. The universe contracts and expands on a regular basis.
   d. Scientists cannot tell if the universe is getting larger or smaller.

______ **2.** Every object in the universe
   a. is not part of any other system.
   b. is part of a larger system.
   c. is scattered in a random pattern.
   d. is part of a structure that is never repeated.

______ **3.** Radiation left over from the big bang is called
   a. cosmic background radiation.
   b. cosmic background noise.
   c. background nuclear radioactivity.
   d. background nuclear energy.

______ **4.** What effect does dark energy seem to have on the universe?
   a. It seems to be accelerating the contraction of the universe.
   b. It seems to be accelerating the expansion of the universe.
   c. It seems to be slowing the expansion of the universe.
   d. It forms about 23% of the total matter of the universe.

______ **5.** The study of the origin, structure, processes, and evolution of the universe is called
   a. the big bang.   **c.** cosmology.
   b. astrology.   **d.** cosmography.

______ **6.** What type of stars do astronomers study to help estimate the age of the universe?
   a. white dwarfs
   b. giants and supergiants
   c. red dwarfs
   d. sunlike stars

______ **7.** After the big bang, what force caused galaxies to form and cluster?
   a. magnetism
   b. fusion
   c. gravity
   d. cosmic background radiation

Assessment

Chapter Test A

Stars, Galaxies, and the Universe
MULTIPLE CHOICE

Write the letter of the correct answer in the space provided.

______ **1.** What is a collection of billions of stars, dust, and gas held together by gravity?
  a. a galaxy
  b. a solar system
  c. a nebula
  d. a cluster

______ **2.** How do astronomers classify galaxies?
  a. by size
  b. by age
  c. by color
  d. by shape

______ **3.** What caused the matter distributed throughout the universe to form galaxies?
  a. gravity
  b. magnetism
  c. cosmic background radiation
  d. parallax

______ **4.** How do scientists tell if a star is cool or warm?
  a. by its size
  b. by its age
  c. by its color
  d. by its shape

______ **5.** What type of galaxy is the Milky Way?
  a. a spiral galaxy
  b. an irregular galaxy
  c. an elliptical galaxy
  d. an odd-shaped galaxy

______ **6.** What kind of stars are shown along the middle of the H-R diagram?
  a. white dwarfs
  b. supergiants
  c. neutron stars
  d. main-sequence stars

_______ **7.** What science studies the origin, structure, processes, and evolution of the universe?
  a. planetology
  b. cosmology
  c. the big bang
  d. astrology

_______ **8.** According to the big bang theory, how did the universe begin?
  a. as a cloud of gases
  b. as a sea of gases
  c. with a small explosion
  d. with a big explosion

_______ **9.** What do scientists observe to find out what early galaxies looked like?
  a. the galaxy Earth is in
  b. galaxies far away in space
  c. the sun
  d. the solar system

_______ **10.** Why is it hard to see the actual motion of stars?
  a. The stars are too close.
  b. The stars are very far away.
  c. The stars move too fast.
  d. The stars move very slowly.

_______ **11.** What process causes energy inside stars to be released?
  a. nuclear fusion
  b. nuclear fission
  c. parallax
  d. apparent magnitude

_______ **12.** What is one force of nature that formed within the first three minutes of the big bang?
  a. gravity
  b. galaxies
  c. matter
  d. energy

Chapter Test A *continued*

MATCHING

Match the correct description with the correct term. Write the letter in the space provided.

_______ **13.** a tight group of stars that looks like a ball

_______ **14.** a large cloud of dust and gas where stars are born

_______ **15.** seems to make the universe expand more quickly

a. nebula

b. globular cluster

c. dark energy

Match the correct description with the correct term. Write the letter in the space provided.

_______ **16.** a star made up of neutrons

_______ **17.** a huge explosion in which a large star dies

_______ **18.** a spinning neutron star that emits pulses of energy

a. neutron star

b. pulsar

c. supernova

FILL-IN-THE-BLANK

Use the terms from the following list to complete the sentences below.

main sequence
light-year

parallax
apparent magnitude

19. In one year, light travels over a distance called a(n)

_______________________.

20. The sun is in the longest stage of its life, called the _______________________.

21. How bright a star is as seen from Earth is called _______________________.

22. The position of a star may seem to shift because of _______________________.

Name _________________________________ Class _______________ Date _______________

Chapter Test B

Stars, Galaxies, and the Universe
MULTIPLE CHOICE
Write the letter of the correct answer in the space provided.

______ **1.** Which of the following types of stars is hottest?
 a. a red giant
 b. a white dwarf
 c. a yellow star
 d. a blue star

______ **2.** The black lines on a star's absorption spectrum indicate
 a. colors of light absorbed by the star's atmosphere.
 b. colors of light emitted by the star's atmosphere.
 c. colors not found in the star's atmosphere.
 d. gases absorbed by the star's atmosphere.

______ **3.** A star appears to shift position when viewed from different locations because of
 a. apparent magnitude.
 b. absolute magnitude.
 c. spectrum.
 d. parallax.

______ **4.** Because stars are so far away,
 a. their actual motion is hard to see.
 b. their actual motion is easy to see.
 c. it is easy to tell that they don't move.
 d. it is hard to tell that they don't move.

______ **5.** Every 24 hours, the stars appear to make a complete circle around Polaris because
 a. Earth's position is fixed.
 b. Earth rotates on its axis.
 c. Polaris rotates on its axis.
 d. the position of Polaris is fixed.

______ **6.** Energy within stars is released in a process called
 a. nuclear fission.
 b. cosmic background radiation.
 c. dark energy.
 d. nuclear fusion.

| Chapter Test B *continued*

_______ **7.** The H-R diagram shows the relationship between a star's absolute
magnitude and its
   a. apparent magnitude.
   b. core temperature.
   c. spectrum.
   d. surface temperature.

_______ **8.** What does a sunlike star become in the final state of its life cycle?
   a. a red giant
   b. a blue giant
   c. a white dwarf
   d. a red dwarf

_______ **9.** Earth is located in a spiral galaxy called
   a. the Andromeda galaxy.
   b. M87.
   c. the Magellanic Cloud.
   d. the Milky Way galaxy.

_______ **10.** What are the three main types of galaxies classified by astronomers?
   a. spiral, triangular, irregular
   b. spiral, elliptical, irregular
   c. spiral, triangular, elliptical
   d. triangular, elliptical, irregular

_______ **11.** A large cloud of gas and dust in interstellar space where stars are born
is called a
   a. supernova.
   b. quasar.
   c. nebula.
   d. globular cluster.

_______ **12.** A collection of billions of stars, dust, and gas bound together by
gravity is a
   a. nebula.
   b. black hole.
   c. galaxy.
   d. planetary system.

_______ **13.** Why do scientists study the light coming from distant galaxies?
   a. to learn what galaxies are made of
   b. to learn about space travel
   c. to learn what early galaxies looked like
   d. to learn about the speed of light

______ **14.** A globular cluster is part of a larger system called a(n)
    a. nebula.
    b. open cluster.
    c. galaxy.
    d. planetary system.

______ **15.** About 14 billion years ago, all matter and energy in the universe was compressed into an extremely small volume that exploded and began expanding in all directions, according to the
    a. big bang theory.
    b. cosmic expansion theory.
    c. unified field theory.
    d. theory of the compression of matter.

______ **16.** Gravity controls the shape and size of the universe because
    a. gravity shifts the position of stars.
    b. gravity compresses matter into a small volume.
    c. gravity acts within dark energy.
    d. gravity acts between objects at great distances.

______ **17.** About how many years after the big bang did the first white dwarf stars form?
    a. one million years
    b. one billion years
    c. 14 billion years
    d. 14 million years

______ **18.** Dark energy seems to be accelerating the expansion of the universe,
    a. compressing the contents of the universe into an extremely small volume.
    b. continuously creating new galaxies.
    c. strengthening the effect of gravity.
    d. counteracting the effect of gravity.

______ **19.** The halos of spiral galaxies contain groups of stars called
    a. globular clusters.
    b. open clusters.
    c. nebulas.
    d. neutron stars.

Chapter Test B *continued*

Use the figure below to answer question 20. Write the letter of the correct answer in the space provided.

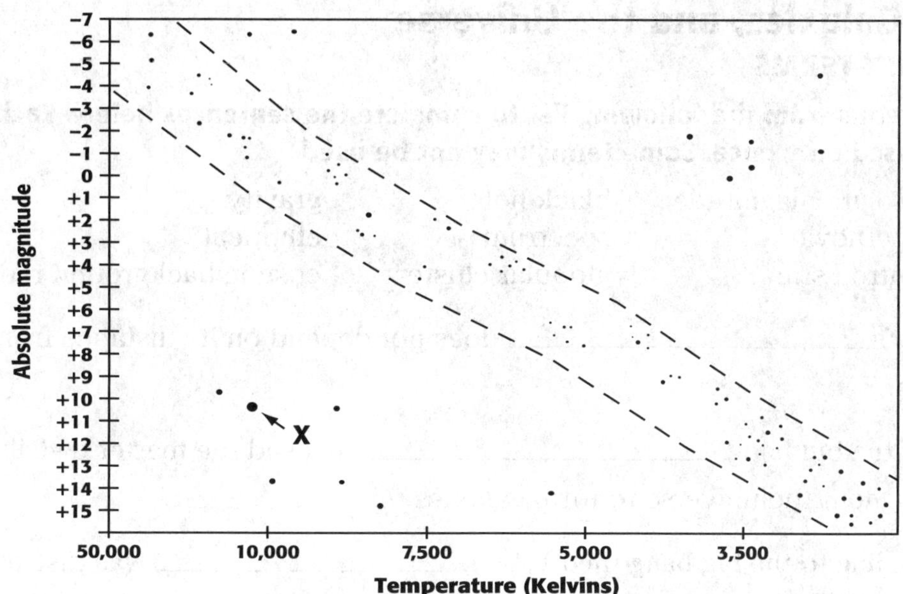

_______ **20.** What type of star might be plotted by the X shown on the H-R diagram above?

  a. a main-sequence star
  b. a red giant
  c. a red dwarf
  d. a white dwarf

MATCHING

Match the correct definition with the correct term. Write the letter in the space provided.

_______ **21.** an object so dense and massive that not even light can escape its gravitational pull

_______ **22.** the band of colors produced when white light passes through a prism

_______ **23.** a gigantic explosion in which a massive star collapses and throws its outer layers into space

_______ **24.** a starlike source of energy in the center of a galaxy

_______ **25.** a spinning neutron star that emits rapid pulses of radio and optical energy

a. supernova

b. pulsar

c. spectrum

d. black hole

e. quasar

Chapter Test C

Stars, Galaxies, and the Universe

USING KEY TERMS

Use the terms from the following list to complete the sentences below. Each term may be used only once. Some terms may not be used.

absolute magnitude	black hole	gravity
supernova	spectrum	elliptical
neutron star	globular cluster	cosmic background radiation

1. A star's _______________________ does not depend on its distance from Earth.

2. After the big bang, _______________________ caused the matter distributed throughout the universe to form galaxies.

3. According to the big bang theory, _______________________ was distributed in every direction as the universe expanded.

4. An object so massive and dense that not even light can escape its gravity is called a(n) _______________________.

5. A highly concentrated group of stars formed at the same time from the same nebula is called a(n) _______________________.

6. Astronomers learn about the composition and temperature of a star by separating the star's light into a(n) _______________________.

UNDERSTANDING KEY IDEAS

Write the letter of the correct answer in the space provided.

_______ **7.** Which of the following magnitudes is brightest?
    a. –1   **c.** –0.11
    b. 0   **d.** +4

_______ **8.** Stars are born in large clouds of gas and dust in interstellar space called
    a. nebulas.   **c.** neutron stars.
    b. galaxies.   **d.** globular clusters.

_______ **9.** A star's apparent shift in position when viewed from different locations is called
    a. apparent magnitude.   **c.** absolute magnitude.
    b. parallax.   **d.** light-year.

| Chapter Test C *continued*

_______ **10.** A hot, dim star that is the leftover center of a red giant is a(n)
    a. red supergiant.   **c.** main-sequence star.
    b. white dwarf.   **d.** blue star.

_______ **11.** According to the big bang theory, the universe was formed about
    a. 200 billion years ago.   **c.** 4.7 billion years ago.
    b. 700 billion years ago.   **d.** 14 billion years ago.

_______ **12.** Stars begin when gravity pulls a ball of gas and dust into a dense
sphere that eventually releases energy in a process called
    a. a supernova.   **c.** nuclear fission.
    b. cosmic radiation.   **d.** nuclear fusion.

_______ **13.** What does the light astronomers observe through telescopes tell them
about distant galaxies?
    a. what the galaxies look like now
    b. what the galaxies looked like in the past
    c. what the galaxies will look like in the future
    d. when the galaxies will disappear

14. Describe the three main types of galaxies classified by their shapes.

____

____

____

15. How can astronomers use absorption spectra of stars to find out which
elements are in a particular star?

____

____

____

____

16. Many scientists believe that the expansion rate of the universe will continue
to increase. If it does, what will happen to the universe?

____

____

____

____

____

CRITICAL THINKING

17. If you traced the paths of the stars in the night sky over a few hours, why would they all appear to circle around Polaris, also called the North Star?

____

____

____

____

18. The sun is about 1.5×10^{11} m from Earth. If the sun suddenly burned out, about how many minutes would elapse before people on Earth would know? (Hint: The speed of light is about 3×10^8 m/s.) Show your work below.

INTERPRETING GRAPHICS

Use the H-R diagram below to answer question 19.

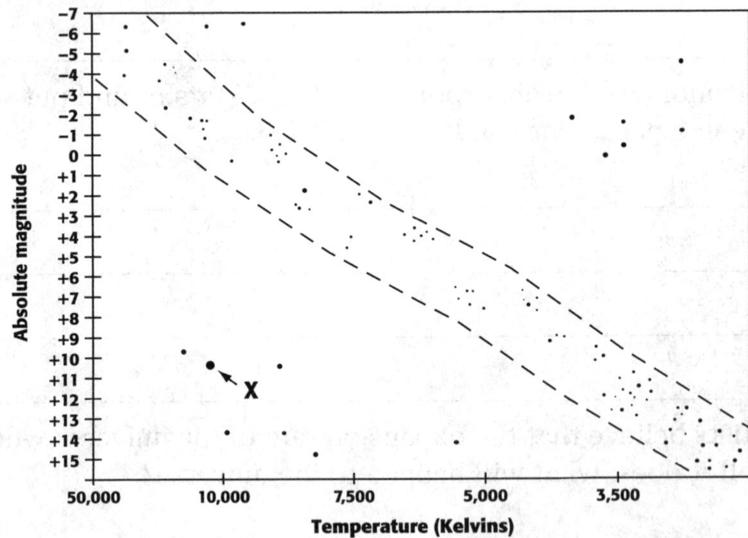

19. Describe the temperature and brightness of the star marked with an X.

____

____

____

____

CONCEPT MAPPING

20. Use the following terms to complete the concept map below:

galaxies	galaxy clusters	globular clusters
nebulas	planets	star clusters

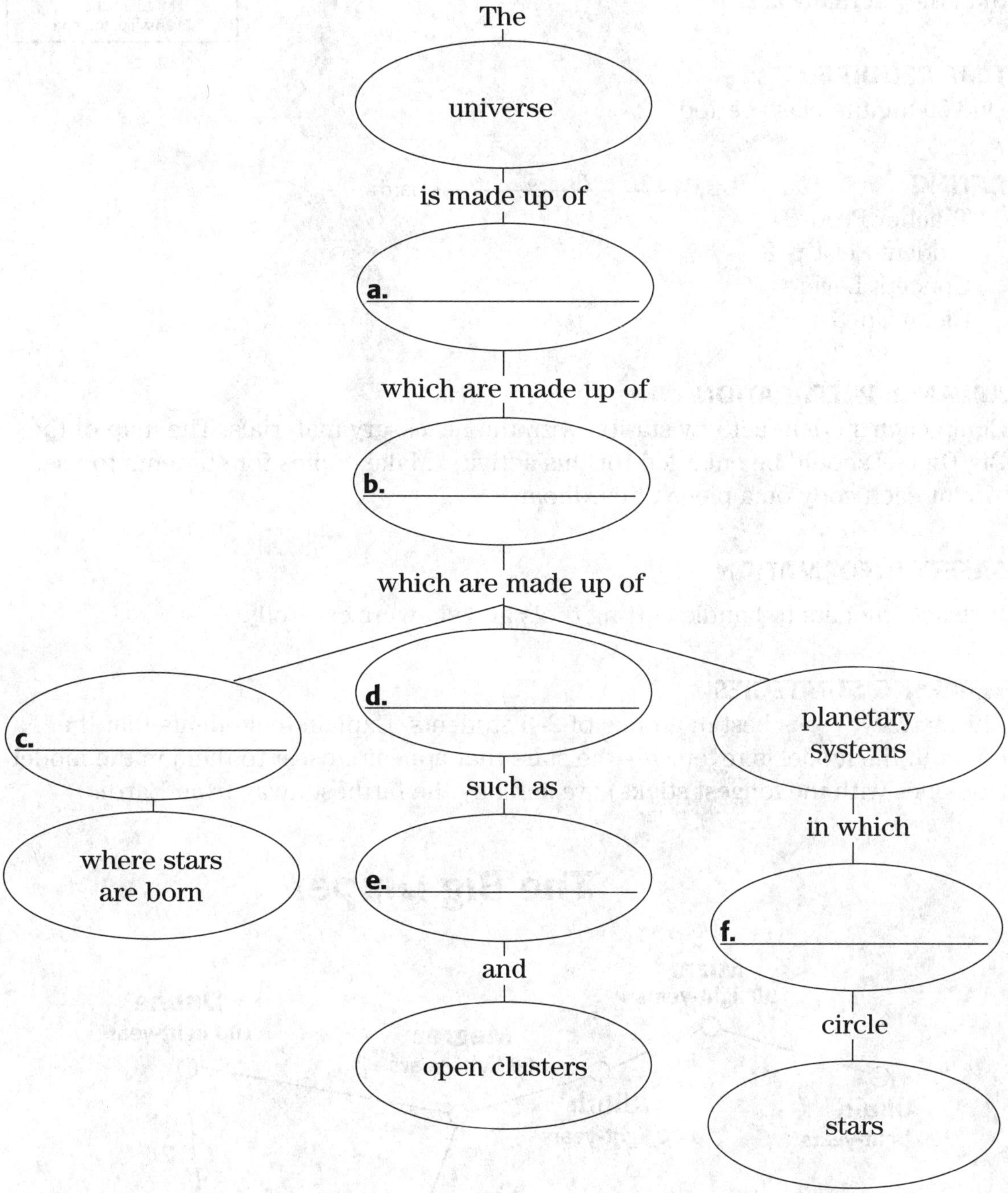

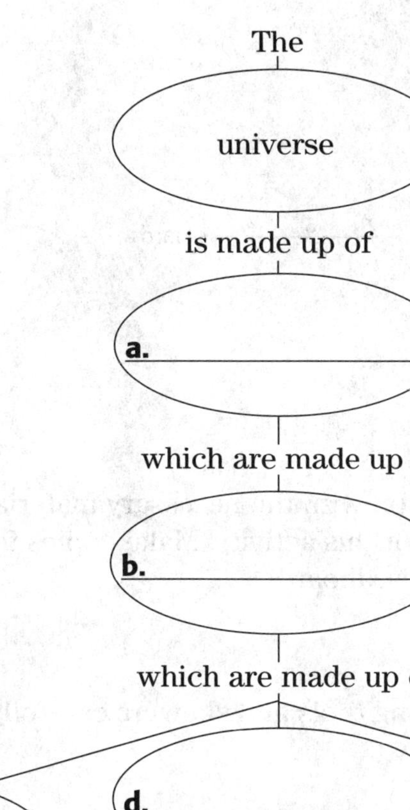

Assessment

Performance-Based Assessment

Teacher Notes

PURPOSE

Students will make a model of the Big Dipper as seen from Earth. They will see that stars appear much closer together than they actually are.

TIME REQUIRED

One 60-minute class period

RATING

Easy ◄— 1 2 3 4 —► Hard

Teacher Prep–3
Student Set-Up–3
Concept Level–3
Clean Up–3

ADVANCE PREPARATION

Equip each student activity station with the necessary materials. The map of the Big Dipper should be enlarged for this activity. Make copies for students to use. Mount each copy on a piece of cardboard.

SAFETY INFORMATION

Instruct students to handle cutting tools and skewers carefully.

TEACHING STRATEGIES

This activity works best in groups of 2–3 students. Explain to students that the effect of the model is reversed—the stars that appear closest to them in the model (the ones with the longest sticks) are actually the farthest away from Earth.

The Big Dipper

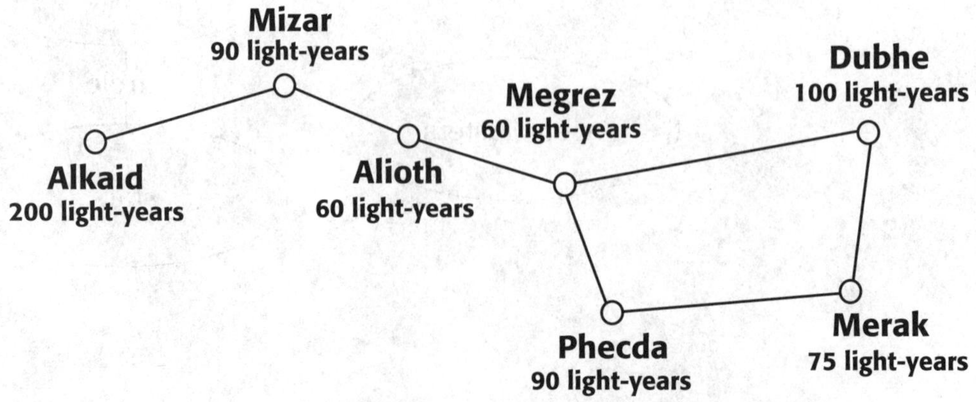

Performance-Based Assessment *continued*

Evaluation Strategies

Use the following rubric to help evaluate student performance.

Rubric for Assessment

Possible points	Appropriate use of materials and equipment (30 points possible)
30–20	Successfully completes activity; safe and careful handling of materials and equipment; close attention to detail; superior lab skills
19–10	Generally completes activity; successful use of materials and equipment; attention to detail; good lab skills
9–1	Does not complete activity; sloppy handling of materials and equipment; no attention to detail; apparent lack of skills
	Quality of Big Dipper model (40 points possible)
40–30	Stars placed accurately; high level of detail
29–20	Stars placed accurately; moderate level of detail
19–10	Stars not placed completely accurately; low level of detail
9–1	Stars placed inaccurately; very low level of detail
	Explanation of observations (30 points possible)
30–20	Clear, detailed explanation; superior knowledge of distances between stars; use of examples to support explanations
19–10	Less clear explanation; adequate understanding of distances in space; minor difficulty in expression
9–1	Unclear explanation; poor understanding of distances in space; substantial factual errors

Name _________________________________ Class _______________ Date __________

Performance-Based Assessment

OBJECTIVE

Although they look close together from Earth, the stars in the Big Dipper are actually very far apart. They also vary tremendously in their distances from Earth. In this activity, you will build a model of the Big Dipper to make these differences clearer.

KNOW THE SCORE!

As you work through the activity, keep in mind that you will be earning a grade for the following:

• how well you work with materials and equipment (30%)

• how well you build your model of the Big Dipper (40%)

• how well you explain your observations (30%)

MATERIALS AND EQUIPMENT

• felt tip pens, blue and red

• map of the Big Dipper mounted on cardboard

• marshmallows, small, 7

• pruning shears, small

• ruler, metric

• skewers, bamboo, 7

SAFETY INFORMATION

PROCEDURE

1. Look at the diagram of the Big Dipper. The distances listed refer to each star's distance from Earth. Convert the distances on the picture into centimeters, using the scale 1 cm = 10 light-years. Cut the bamboo skewers to these lengths. How long should the skewers be to represent the following distances?

   50 light-years: ________________________

   60 light-years: ________________________

   75 light-years: ________________________

   90 light-years: ________________________

   100 lightyears: ________________________

   200 light-years: ________________________

2. Put a small clump of clay on each star on the map, and stick the bamboo skewers into the clay with the correct length skewer at each star. The skewers should be pushed through the clay, the map, and the cardboard so that the tips of the skewers show on the reverse side of the cardboard. Stick a marshmallow on the end of each bamboo skewer.

Performance-Based Assessment *continued*

3. The star known as Alkaid is a blue star. Use the blue felt tip pen to mark the marshmallow representing Alkaid. This star is farthest from Earth.

4. The star known as Dubhe is a red giant. Use the red felt tip pen to mark the marshmallow representing Dubhe.

ANALYSIS

5. According to your diagram of the Big Dipper, which star appears nearest to Alkaid when viewed from Earth?

____

6. What two factors affect a star's apparent magnitude?

____

____

7. If all the stars in this constellation had the same absolute magnitude, which would appear brightest to people on Earth?

____

8. What is a light-year?

____

____

9. It takes about 8 minutes for light from the sun to reach Earth. Using the definition of a light-year as a model, how could you describe the distance of the sun from Earth?

____

____

BIG IDEA QUESTION

10. Is the star you named in item 5 necessarily the closest star in the Big Dipper to Alkaid? Explain your answer.

____

____

____

____

Assessment

Standards Assessment

Teacher Notes and Answer Key

To provide practice under more realistic testing conditions, give students 20 min to answer all of the questions in this assessment.

QUESTION NUMBER	CORRECT ANSWER	STANDARD
1	A	8.2.g (supporting)
2	C	
3	A	8.4.b (supporting)
4	D	8.2.g (supporting)
5	C	8.4.c (supporting)
6	D	8.4.b (mastering)
7	B	8.4.b (mastering)
8	A	8.4.a (mastering)
9	D	8.4.b (exceeding)
10	B	8.2.g (mastering)
11	A	8.3.a (mastering)
12	A	8.7.a (mastering)
13	D	7.4.d (mastering)
14	B	7.4.g (exceeding)

TEST DOCTOR

The following Standards Assessment questions have been diagnosed by the Test Doctor. Find out what might be causing your students' "ailing" answers. Each Test Doctor is followed by a diagnostic teaching tip to help you address students' learning needs.

Question 1 *asks students to identify the definition of* structure.

A Correct. The noun form of structure refers to the arrangement of parts in a whole.

B Incorrect. *Identity*, the condition of being a specific person or thing, is not the correct word choice.

C Incorrect. *Concept*, an idea or thought, is not the correct word choice.

D Incorrect. *Cycle*, a complete set of events recurring in the same sequence, is not the correct word choice.

Diagnostic Teaching Tip: Students who have difficulty with this question might be confusing the noun and verb forms of the word structure. Discuss how a noun structure, the school for example, differs from the verb structure. Use several blocks to structure (verb) a pyramid. Point out that the pyramid is a structure (noun).

Question 2 *asks students to choose the correct form of the verb* maintain *in context.*

A Incorrect. *Maintaining* can be used as a verb but only with the help of the auxiliary verb *to be* (was maintaining). This form does not fit this sentence.

Standards Assessment *continued*

B Incorrect. *Maintains* is the present tense. Because the verb in the subordinate clause is in the past tense, the verb in the main clause should also be in the past tense.

C Correct. The past tense, *maintained*, is the correct form of the verb for this sentence.

D Incorrect. *Maintenance* is a noun. A verb is required to complete the sentence.

Diagnostic Teaching Tip: Students who struggle with this type of question might have misread the sentence. Encourage students to read the sentence aloud, filling in the blank with each answer choice. Usually, the correct answer will "sound right" when read aloud. If the student is still having trouble, they may benefit from creating a chart showing the tenses of several regular verbs.

Question 3 *asks students to demonstrate understanding of the word* evolution *used in the context of nonliving matter.*

A Correct. *Evolution* in this sentence refers to the patterns of change and growth in the movements, characteristics, and life cycles of stars.

B Incorrect. Since stars are not living organisms, *evolution* in this sense does not apply to them.

C Incorrect. *Heredity* describes the way characteristics are passed along in organisms; additionally, stars are not living organisms.

D Incorrect. Since we have not yet discovered life elsewhere in the universe, *evolution* in this sense does not apply to this sentence.

Diagnostic Teaching Tip: Students who choose incorrect answers for this question might benefit from a discussion comparing and contrasting evolution in living organisms and in nonliving things.

Question 4 *asks students to define the word* role *without context.*

A Incorrect. *Spin,* to rotate swiftly, is not synonymous with role.

B Incorrect. *Personality,* the quality of being a person or individuality, does not mean the same thing as role.

C Incorrect. *Tumble,* to do somersaults, handsprings, or similar acrobatic feats, is not synonymous with role.

D Correct. *Function* (the normal or characteristic action of anything) and role (a function or office assumed by someone) are close in meaning.

Diagnostic Teaching Tip: Students who answer this question incorrectly might be confusing role and roll. Answers A and C, spin and tumble, are both synonyms for different aspects of roll. Students who choose answer B might be thinking of a role an actor would play, although role and personality are not synonymous.

Question 5 *asks students to demonstrate an understanding of the concept of the light-year.*

A Incorrect. Apparent magnitude is the brightness of a star as seen from Earth.

Standards Assessment *continued*

B Incorrect. Absolute magnitude is the brightness that a star would have at 32.6 light-years from Earth.

C Correct. A light year is the distance that light travels in a year.

D Incorrect. Parallax is the apparent shift in the position of an object when the object is viewed from different locations.

Diagnostic Teaching Tip: Students who struggle to answer this question might benefit from reviewing the relationship between apparent magnitude, absolute magnitude, parallax, and light-year. Create a data sheet for a star that includes the star's distance from Earth in kilometers and its apparent magnitude. Have students calculate its absolute magnitude and distance from Earth in light years.

Question 6 *asks students to demonstrate an understanding of how stars differ in color and temperature.*

A Incorrect. Blue-white stars are very hot, with a surface temperature of 10,000° C to 30,000 degrees° C.

B Incorrect. Yellow stars are not the coolest of these stars. They have a surface temperature of 5,000° C to 6,000° C.

C Incorrect. Yellow-white stars are hotter than yellow stars, with a surface temperature of 6,000° C to 7,500° C.

D Correct. Orange stars are the coolest stars listed here. They have a surface temperature of 3,500° C to 5,000° C. The only stars cooler than orange stars are red stars.

Diagnostic Teaching Tip: Students who have difficulty answering this question correctly might benefit from using a mnemonic device to help them remember the relative temperature of different stars. For example, Oh Boy A Fine Girl/Guy Kissed Me or Ovens Bake A Fairly Good Ketchup Mousse. Have the students create their own mnemonic.

Question 7 *asks students to demonstrate understanding of the methods by which scientists determine the characteristics of stars.*

A Incorrect. The brightness of a star, absolute magnitude, is calculated using its apparent magnitude and its distance from Earth.

B Correct. Each element in a star produces a unique pattern of emission lines, allowing the scientists to determine what gases the star contains and the stage it occupies in its life cycle.

C Incorrect. Scientists use parallax and trigonometry to determine a star's distance from Earth, which is measured in light years.

D Incorrect. Scientists use a combination of factors to determine the size of a star.

Diagnostic Teaching Tip: Students who struggle to answer this question might benefit from a review of the definition of emission lines and how they affect a star's absorption pattern. Ask students to write a paragraph explaining how the absorption pattern for an element like neon (Ne) might be different from the absorption pattern for a star.

Standards Assessment *continued*

Question 8 *asks students to demonstrate understanding of the differences between types of galaxies.*

A Correct. This round galaxy is called an *elliptical galaxy*. About one-third of all galaxies are elliptical—they can appear as spheres or more elongated.

B Incorrect. This is a galaxy, but not an *irregular galaxy*. An irregular galaxy would be oddly shaped and not easily classified as spiral or elliptical.

C Incorrect. A *supernova* is the explosion of a massive star. These stars are still clustered together.

D Incorrect. This is a galaxy, but not a *spiral galaxy*. A spiral galaxy has a bulge in the center and spiral arms.

Diagnostic Teaching Tip: Students who struggle to answer this type of question might benefit from a review of the characteristics of the different types of galaxies.

Question 9 *asks students to demonstrate understanding that stars may differ in size, temperature and color*

A Incorrect. Alpha Centauri is brighter than the sun, but it is not hotter.

B Incorrect. Sadr is both hotter and brighter than Pollux.

C Incorrect. Pollux is the coolest star shown in the graph.

D Correct. Because Sadr is located both above and to the left of the other stars in the graph, we know that it is both the hottest and the brightest star.

Diagnostic Teaching Tip: Students who have difficulty answering this question correctly might have become confused because we are used to graphs in which amounts increase as we move from left to right, and in which intensity increases as we move from negative to positive. Discuss why scientists have developed a system that reads so much differently than other systems we are used to reading. Ask students to take sides to debate the pros and cons of such a system.

Question 10 *asks students to demonstrate understanding of the role gravity plays in shaping the universe.*

A Incorrect. Dark energy, not gravity, seems to be accelerating the expansion of the universe.

B Correct. Gravity organizes the matter in space into visible patterns such as nebulas, galaxies, and planetary systems.

C Incorrect. Gravity does attract bodies to one another, but they don't necessarily have similar compositions.

D Incorrect. These bodies are not scattered randomly throughout the universe but are organized into patterns due to the mutual attraction of gravity between bodies.

Standards Assessment *continued*

Diagnostic Teaching Tip: Students who have difficulty answering this question correctly might benefit from reviewing the basic properties of gravity and inertia and their combined effects on bodies in space. Have students draw diagrams of the moon orbiting Earth. Ask them to mark with arrows the directions of the opposing forces at work on the moon. Discuss why the moon's path around Earth is roughly circular.

Question 11 *asks students to demonstrate an understanding of the essential features of an atom.*

A Correct. The simplest atom consists of just one proton and one electron; in that case, the proton is the nucleus.

B Incorrect. Neutrons do not form the basis for atomic nuclei; protons do.

C Incorrect. All atoms have electrons, but electrons are not found in nuclei.

D Incorrect. A positron is an antiparticle and not found in the atomic nuclei.

Diagnostic Teaching Tip: Students who have difficulty answering this question might benefit from a review of atomic structure, particularly with regard to what makes an atom the element that it is.

Question 12 *asks students to identify regions of the periodic table.*

A Correct. Helium is found in Group 18, the noble gases.

B Incorrect. Group 13 contains metalloids and metals, not gases like helium.

C Incorrect. Group 15, the nitrogen group, contains metals, metalloids, and nonmetals, but they are all solid at room temperature. Helium is a gas at room temperature.

D Incorrect. Group 1 contains only metallic elements. Helium is a gas.

Diagnostic Teaching Tip: Students who have difficulty answering this question correctly might benefit from reviewing placement and shared properties of elements in the periodic table. Create a trivia game in which teams of students compete to locate elements and groups of elements in the periodic table and identify their properties.

Question 13 *asks students to demonstrate understanding that radioactive dating indicates that Earth is approximately 4.6 billion years old.*

A Incorrect. *Erosion* is the process by which wind and water transport soil and sediment from one location to another.

B Incorrect. The *fossil record* mainly represents the Phanerozoic eon, which is the eon in which we live.

C Incorrect. Because no one has ever seen what the *mantle* really looks like, scientists must infer what the characteristics of the mantle are from observations they make on Earth's surface.

D Correct. Scientists believe that the moon is the same age as Earth, so they date rocks and meteors from the moon using *radioactive dating* to hypothesize how old Earth is.

Standards Assessment *continued*

Diagnostic Teaching Tip: Students who answer this question incorrectly might benefit from a review of the methods and sources of geological dating. How is such dating performed? What have scientists learned about Earth's geological history by dating fossils, cave paintings, and other artifacts? What role does radioactive dating play?

Question 14 *asks students to explain significant developments of Earth's life forms on a geological time scale.*

A Incorrect. Cyanobacteria were one of the first organisms to appear on Earth. They obtained energy through photosynthesis.

B Correct. Before the appearance of cyanobacteria, Earth's atmosphere was unsuitable for life as we know it today.

C Incorrect. Cyanobacteria are producers.

D Incorrect. Oxygen produced by the cyanobacteria formed the ozone layer that protects Earth from the sun's radiation.

Diagnostic Teaching Tip: Answering this question correctly depends upon knowing the role of cyanobacteria in Earth's history. Ask students to research the role of anaerobic organisms in Earth's geologic history. Do they still exist? What role do they play?

Standards Assessment

REVIEWING ACADEMIC VOCABULARY

_______ **1.** Which of the following words means "a whole that is built or put
together from parts"?
    A structure
    B identity
    C concept
    D cycle

_______ **2.** Choose the appropriate form of the word *maintain* for the following
sentence: "She _______ a B average at school, even while she was in the
play."
    A maintaining
    B maintains
    C maintained
    D maintenance

_______ **3.** In the sentence "Scientists study the universe by studying the
evolution of stars," what does the word *evolution* mean?
    A the pattern of change and growth
    B the process by which species develop
    C the way characteristics are passed along
    D the advancement of life in space

_______ **4.** Which of the following words is the closest in meaning to the
word *role*?
    A spin
    B personality
    C tumble
    D function

REVIEWING CONCEPTS

_______ **5.** What is the unit that astronomers use to measure the distances
between Earth and stars called?
    A apparent magnitude
    B absolute magnitude
    C light-year
    D parallax

_______ **6.** Which of the following stars has the coolest temperature?
    A a blue-white star
    B a yellow star
    C a yellow-white star
    D an orange star

_______ **7.** What do scientists learn by studying the pattern of lines in a star's absorption spectrum?
A the brightness of the star
B the elements of the star
C the distance of the star from Earth
D the size of the star

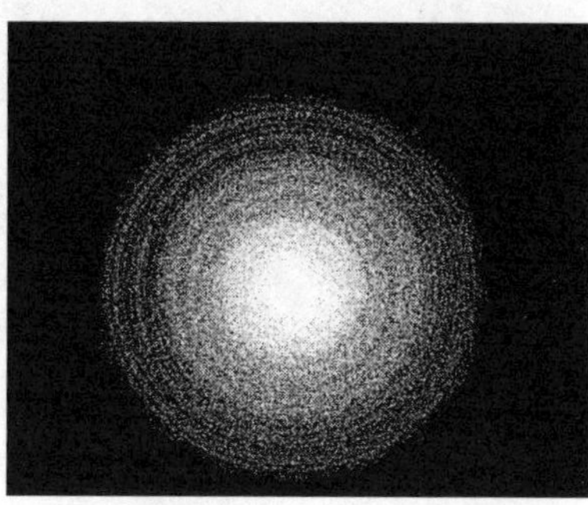

_______ **8.** The picture above shows an example of
A an elliptical galaxy.
B an irregular galaxy.
C a supernova.
D a spiral galaxy.

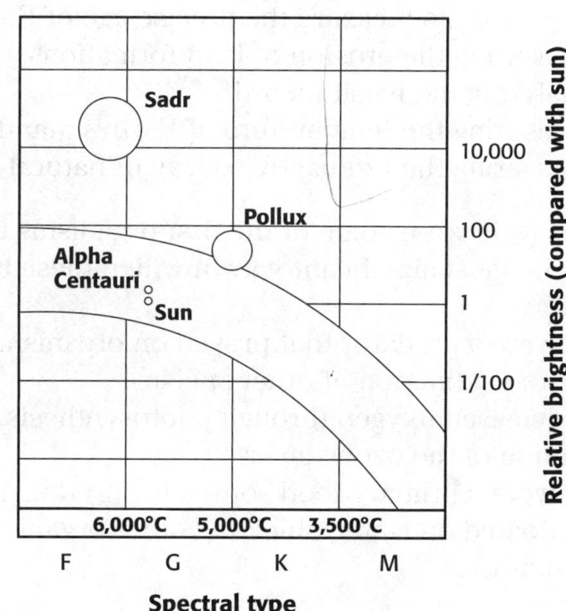

_______ **9.** Which statement about the H-R diagram above is true?
A Alpha Centauri is hotter and brighter than the sun.
B Sadr is cooler but brighter than the star Pollux.
C Pollux is the hottest star shown in the graph.
D Sadr is the hottest star shown in the graph.

❚ Standards Assessment *continued*

_______ **10.** The universe contains galaxies, stars, and planets. How does gravity affect these bodies in space?
    A Gravity pulls bodies away from each other.
    B Gravity organizes bodies into nebulas, galaxies, and planetary systems.
    C Gravity attracts bodies with similar compositions to each other.
    D Gravity causes bodies to be scattered randomly throughout the universe.

REVIEWING PRIOR LEARNING

_______ **11.** Which of the following particles is always found in the nucleus of an atom?
    A protons
    B neutrons
    C electrons
    D positrons

_______ **12.** Which group in the periodic table contains helium?
    A Group 18, the noble gases
    B Group 13, the boron group
    C Group 15, the nitrogen group
    D Group 1, the alkali metals

_______ **13.** How do geologists measure the precise age of Earth?
    A by measuring the erosion of land formations
    B by analyzing the fossil record
    C by measuring the temperature of Earth's mantle
    D by measuring the radioactive decay of natural elements

_______ **14.** Cyanobacteria were some of the first organisms to appear on Earth. What is the most significant way in which these bacteria affected life on Earth?
    A They were a predator that preyed on organisms that threatened to cause the extinction of other species.
    B They released oxygen through photosynthesis, which led to the formation of the ozone layer.
    C They were a primary food source for producers.
    D They floated on water, shielding other organisms from the sun's radiation.

DATASHEET

Exploring the Movement of Galaxies in the Universe

Teacher Notes

In this activity, students will use a stretched rubber band to model the expansion of the universe (covers standard 8.4.a). Do not allow students to play carelessly with their rubber bands.

MATERIALS

For each group

- marker
- rubber band, thick
- ruler, metric
- scissors

SAFETY CAUTION

Remind students to review all safety cautions and icons before beginning this activity. Make sure that students wear their goggles when stretching the rubber bands.

Exploring the Movement of Galaxies in the Universe

In this activity, you will explore why distant galaxies appear to be moving away from the Milky Way galaxy faster than galaxies close to the Milky Way galaxy are.

SAFETY INFORMATION

PROCEDURE

1. Use **scissors** to cut a thick **rubber band** so that it is at least 4 cm long.
2. Get a **metric ruler** and a **marker.**
   - On the rubber band, draw five ovals that are 1 cm apart from each other.
   - Label the ovals "A" through "E" from left to right.
   - Oval A stands for the Milky Way galaxy.
   - Ovals B through E stand for galaxies at different distances from the Milky Way galaxy.
3. Stretch the rubber band to double its starting length.
   - The ovals should 2 cm apart now.
4. Stretch the rubber band again—if possible—so that it is three times its starting length.
   - The ovals should be 3 cm apart now.

ANALYSIS

5. Look at the increase in distance between "galaxies" A and B on the rubber band. How does it compare to the increase in distance between B and C?

____

____

____

____

____

6. Using your rubber band as an example, explain why distant galaxies appear to move away faster than closer ones.

____

____

____

Name ________________________________ Class ______________ Date ____________

Exploring the Movement of Galaxies in the Universe

In this activity, you will explore why distant galaxies appear to be moving away from the Milky Way galaxy faster than galaxies close to the Milky Way galaxy are.

SAFETY INFORMATION

PROCEDURE

1. Using **scissors**, cut a thick **rubber band** so that it is at least 4 cm long.

2. Using a **metric ruler** to obtain accurate measurements and a **marker,** draw five ovals 1 cm apart from each other on the rubber band. Label the ovals "A" through "E" from left to right. Oval A represents the Milky Way galaxy, and ovals B through E represent galaxies at various distances from the Milky Way galaxy.

3. Stretch the rubber band to double its original length so that the ovals are 2 cm apart.

4. Stretch the rubber band again—if possible—so that it is triple its original length and the ovals are 3 cm apart.

ANALYSIS

5. How did the distances between galaxies increase as the rubber band was stretched?

____

____

____

____

____

6. Using your rubber band as an example, explain why distant galaxies appear to move away faster than closer ones.

____

____

____

____

____

____

　　　　　　　　　　　　　　　　　　　　　DATASHEET C

Exploring the Movement of Galaxies in the Universe

In this activity, you will explore why distant galaxies appear to be moving away from the Milky Way galaxy faster than galaxies close to the Milky Way galaxy are.

SAFETY INFORMATION

PROCEDURE

1. Using **scissors,** cut a thick **rubber band** so that it is at least 4 cm long.

2. Using a **metric ruler** to obtain accurate measurements and a **marker,** draw five ovals 1 cm apart from each other on the rubber band. Label the ovals "A" through "E" from left to right. Oval A represents the Milky Way galaxy, and ovals B through E represent galaxies at various distances from the Milky Way galaxy.

3. Stretch the rubber band to double its original length so that the ovals are 2 cm apart.

4. Stretch the rubber band again—if possible—so that it is triple its original length and the ovals are 3 cm apart.

ANALYSIS

5. What did the model show about the distance between any two neighboring galaxies?

____

____

____

____

6. Using your rubber band as an example, explain why distant galaxies appear to move away faster than closer ones.

____

____

____

____

Demonstrating Parallax

Teacher Notes

This activity has students model parallax or the apparent shift in the position of
an object when viewed from different locations (covers standard 8.4.c).

SAFETY CAUTION

Remind students to review all safety cautions and icons before beginning
this activity.

Quick Lab   **DATASHEET A**

Demonstrating Parallax

PROCEDURE

1. Hold your thumb in front of your face at arm's length.

2. Close your right eye.
   - Find an object some distance away.
   - Use your thumb to block out the object.

3. Open your eye and close the left eye.
   - Do not move your head.
   - Look at the object that you covered with your thumb in step 2.
   - What side of your thumb is the object on now?

____

____

____

____

____

4. Now, bring your thumb close to your face.
   - Repeat steps 2 and 3.
   - Did the object seem to move about the same distance as it did in step 3, or farther?

____

____

____

____

____

5. Parallax explains how the positions of stars seem to shift when we look at them from Earth.
   - How does this activity show how parallax works?

____

____

____

____

____

Demonstrating Parallax

PROCEDURE

1. Hold your thumb in front of your face at arm's length.

2. Close one eye, and use your thumb to cover an object some distance away.

3. Without moving your head, open your eye and close the other eye. What seems to happen to your thumb relative to the object that you covered in step 2?

____

____

____

____

____

____

____

____

____

4. Now, bring your thumb close to your face, and repeat steps 2 and 3.

5. How does this activity demonstrate parallax?

____

____

____

____

____

____

____

____

                                        DATASHEET C

Demonstrating Parallax

PROCEDURE

1. Hold your thumb in front of your face at arm's length.

2. Close one eye, and use your thumb to cover an object some distance away.

3. Without moving your head, open your eye and close the other eye. Describe your observations.

____

____

____

____

4. Now, bring your thumb close to your face, and repeat steps 2 and 3. Describe your observations.

____

____

____

____

____

5. How does this activity demonstrate parallax?

____

____

____

____

____

____

____

Making a Star Movie

Teacher Notes

This activity has students model the life cycle of a star (covers standard 8.4.b).

MATERIALS

For each student

- markers or pencils, colored (set)
- notebook, spiral

SAFETY CAUTION

Remind students to review all safety cautions and icons before beginning this activity.

Making a Star Movie

A single star can appear very different at different stages in its life. One way to model this life cycle is to make a flip-chart movie that shows the process.

TRY IT!

1. Draw the stages in the life cycle of a star.
- Use a **marker** or **colored pencils**.
- Draw pictures in the lower right-hand corners of the pages of a **spiral notebook.**
- Use the correct colors for each stage of the star's life.
- Start on the last page of the notebook.
- For example, draw a cloud of gas and dust on the last page; draw a main-sequence star on the page before the last page; draw a red giant or red supergiant on the page before that; and draw a white dwarf star on the page before that.

2. Use as many pages as you need.
- You can draw two to four pictures for each stage of a star's life.
- Remember that some stages last longer than others. Give these stages more pages.
- You will have between ten and sixteen pictures.

3. If you want, you may read the rest of the section to learn about other stages of stars.
- To show the end of a star's life, you may choose to draw a supernova, a neutron star, a pulsar, or a black hole.

4. When you have finished your drawings, take the notebook in your left hand.
- Hold the corner of the notebook with the thumb and first finger of your right hand.
- Flip the pages so that they pass before your eyes like the pictures on a movie screen.

THINK ABOUT IT!

5. Which stage of a star's life cycle did you use the most pages to draw?

____

6. Compare your drawings.
- During which stage of its life cycle was the star the largest and brightest?

____

- During which stage of its life cycle was the star the smallest and dimmest?

Quick Lab   **DATASHEET B**

Making a Star Movie

A single star can appear very different at different stages in its life. One way to model this life cycle is to make a flip-chart movie that shows the process.

TRY IT!

1. Using a **marker** or **colored pencils,** draw the stages in the life cycle of a star on the corners of the pages of a **spiral notebook.** Be sure to match the color of the star with the star's correct stage of life.

2. Use as many pages as necessary to illustrate each stage in the star's life. The star will form from a cloud of gas and dust, will become a main-sequence star, will change into a red giant or red supergiant, and eventually will become a white dwarf.

3. If you want, you may read the rest of the section and choose to draw a supernova, a neutron star, a pulsar, or a black hole.

4. When you have completed your illustrations, take the notebook in your left hand. Using the thumb and first finger of your right hand, flip the pages so that they pass before your eyes like the images on a movie screen.

THINK ABOUT IT!

5. Which stage of a star's life cycle did you use the most pages to illustrate?

____

____

6. During which stage of its life cycle was the star the largest? During which stage was the star the smallest? During which stage was the star the brightest? Finally, during which stage was the star the dimmest?

____

____

____

____

____

Quick Lab   **DATASHEET C**

Making a Star Movie

A single star can appear very different at different stages in its life. One way to
model this life cycle is to make a flip-chart movie that shows the process.

TRY IT!

1. Using a **marker** or **colored pencils,** illustrate the stages in the life cycle of a
   star on the corners of the pages of a **spiral notebook.** Be sure to use
   appropriate colors.

2. Use as many pages as necessary to illustrate each stage in the star's life.

3. Read the rest of the section. Illustrate a supernova, a neutron star, a pulsar, or
   a black hole as a stage in the star's life.

4. When you have completed your illustrations, take the notebook in your left
   hand. Using the thumb and first finger of your right hand, flip the pages so
   that they pass before your eyes like the images on a movie screen.

THINK ABOUT IT!

5. Which stage of a star's life cycle did you use the most pages to illustrate?

   ____

   ____

6. Identify the stages in its life cycle where the star is the largest, smallest,
   brightest, and dimmest.

   ____

   ____

   ____

   ____

   ____

   ____

Quick Lab | **DATASHEET**

Modeling Galaxies

Teacher Notes

This activity has students model a spiral galaxy (covers standard 8.4.a).

MATERIALS

For each student

- glitter, 1 tsp
- mixing bowl
- spoon (optional)
- water

SAFETY CAUTION

Remind students to review all safety cautions and icons before beginning this activity.

Quick Lab

Modeling Galaxies

SAFETY INFORMATION

PROCEDURE

1. Fill a **medium mixing bowl** with **water.**
   - Fill the bowl two-thirds full.

2. Place **1 tsp of colored glitter** in the center of the bowl.

3. Use your fingers or a **large stirring spoon** to stir the water.
   - Stir the water quickly.

4. Does the glitter form a pattern?
   - What shape does the glitter make?

   ____

5. What type of galaxy does the glitter model?

   ____

   ____

Quick Lab   **DATASHEET B**

Modeling Galaxies

SAFETY INFORMATION

PROCEDURE

1. Fill a **medium mixing bowl** two-thirds full of water.
2. Place **1 tsp of colored glitter** in the center of the bowl.
3. Use your fingers or a **large stirring spoon** to rapidly swirl the water in the bowl.
4. What kind of pattern does the glitter form?

____

____

____

____

5. What type of galaxy does the glitter model?

____

____

____

____

____

Modeling Galaxies

DATASHEET C

SAFETY INFORMATION

PROCEDURE

1. Fill a **medium mixing bowl** two-thirds full of **water.**

2. Place **1 tsp of colored glitter** in the center of the bowl.

3. Use your fingers or a **large stirring spoon** to rapidly swirl the water in the bowl.

4. What observations can you make about the glitter?

   ____

5. What type of galaxy have you modeled?

   ____

    DATASHEET

The Expanding Universe

Teacher Notes

This activity has students demonstrate that all objects in the universe are moving away from each other (covers standard 8.2.g).

SAFETY CAUTION

Remind students to review all safety cautions and icons before beginning this activity.

MATERIALS

For each group

- balloon
- marker
- ruler, metric
- string

The Expanding Universe

Use a balloon as a model to show that all objects in the universe are moving away from each other.

SAFETY INFORMATION

TRY IT!

1. Use a **marker** to make three dots in a row on a **balloon** that has no air in it.
   - Label the dots "A," "B," and "C." Dot B should be closer to dot A than to dot C.

2. Blow up the balloon just until it holds its shape.
   - Pinch the balloon to keep it blown up, but do not tie the neck.

3. Get **string** and a **metric ruler.**
   - Put one end of the string on dot A and run the string straight to dot B. Pinch the string where it meets dot B.
   - Measure the string from the end to the pinched spot. This is the distance

     from A to B: _____________________

   - Repeat to measure the distance from B to C: _____________________

   - Repeat to measure the distance from A to C: _____________________

4. With the balloon still blown up, blow more air into it until it is twice as big.

The Expanding Universe *continued*

5. Use the string and ruler to measure the distances again. Write them down.

More air, A to B: _______________________

More air, B to C: _______________________

More air, A to C: _______________________

- For each pair of dots, subtract the distances you wrote in step 3 from the distances you found in this step. Then, divide by 2. This math will give you the rate of change for each pair of dots. Show your work below.

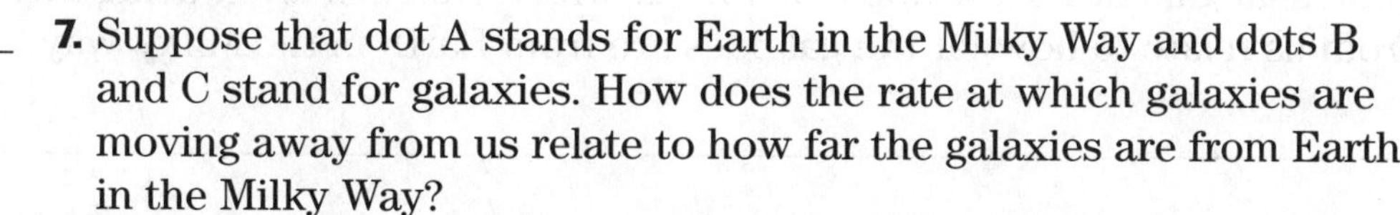

____________ − ____________ = ____________ ÷ 2 = ____________

More air,   1st A to B   A to B rate

A to B   of change

____________ − ____________ = ____________ ÷ 2 = ____________

More air,   1st B to C   B to C rate

B to C   of change

____________ − ____________ = ____________ ÷ 2 = ____________

More air,   1st A to C   A to C rate

A to C   of change

THINK ABOUT IT!

6. Did the difference between dots A and B, B and C, or A and C show the greatest rate of change?

____

______   **7.** Suppose that dot A stands for Earth in the Milky Way and dots B and C stand for galaxies. How does the rate at which galaxies are moving away from us relate to how far the galaxies are from Earth in the Milky Way?
      a. The closer galaxies are not moving away at all.
      b. The middle-distance galaxies move at the slowest rate.
      c. The farther away galaxies move away at a faster rate.

The Expanding Universe

Use a balloon as a model to show that all objects in the universe are moving away from each other.

SAFETY INFORMATION

TRY IT!

1. Use a **marker** to make three dots in a row on a noninflated **balloon.** Label the dots "A," "B," and "C." Dot B should be closer to dot A than to dot C.

2. Blow up the balloon just until it is taut. Pinch the balloon to keep it inflated, but do not tie the neck.

3. Use **string** and a **metric ruler** to measure the distances between dots A and B, B and C, and A and C.

4. With the balloon still inflated, blow into the balloon until its diameter is twice as large.

5. Measure the distances between dots A and B, B and C, and A and C. For each pair of dots, subtract the original distances measured in step 3 from the new distances. Then, divide by 2. This calculation will give you the rate of change for each pair of dots.

THINK ABOUT IT!

6. Which pair of dots experienced the greatest rate of change?

____

7. Suppose that dot A represents Earth in the Milky Way and dots B and C represent galaxies. How does the rate at which galaxies are moving away from us relate to how far the galaxies are from Earth in the Milky Way?

____

____

____

____

                                            DATASHEET C

The Expanding Universe

Use a balloon as a model to show that all objects in the universe are moving away from each other.

SAFETY INFORMATION

TRY IT!

1. Use a **marker** to make three dots in a row on a noninflated **balloon.** Label the dots "A," "B," and "C." Dot B should be closer to dot A than to dot C.

2. Blow up the balloon just until it is taut. Pinch the balloon to keep it inflated, but do not tie the neck.

3. Use **string** and a **metric ruler** to measure the distances between dots A and B, B and C, and A and C.

4. With the balloon still inflated, blow into the balloon until its diameter is twice as large.

5. Measure the distances between dots A and B, B and C, and A and C. Calculate the rate of change for each pair of dots. Show your work below.

THINK ABOUT IT!

6. Which pair of dots showed the greatest rate of change?

____

7. Suppose that dot A represents Earth in the Milky Way and dots B and C represent galaxies. Based on your observations in this activity, explain the relationship between the distances of the galaxies from Earth in the Milky Way and the rate at which the galaxies are moving away from us.

____

____

____

____

DATASHEET

Star Colors: Red Hot, or Not?

Teacher Notes

This lab has students experiment with a light bulb and batteries to learn what the color of a glowing object reveals about the temperature of the object. The experiment will help students to discover how the color of stars enables astronomers to estimate the approximate temperature of stars (covers standards 8.4.a and 8.9.a).

TIME REQUIRED

One 45-minute class period

LAB RATINGS

Easy ← 1　2　3　4 → Hard

Teacher Prep—2
Student Set-Up—2
Concept Level—1
Clean Up—1

MATERIALS

The materials listed on this page are enough for a group of three or four students.

SAFETY CAUTION

Remind students to review all safety cautions and icons before beginning this lab activity. Be sure that the students disconnect the wires at each step. If left connected, the wires can get very hot.

LAB NOTES

Students may find that holding the wires to the light bulb is difficult. If so, you may substitute a light socket. Any miniature incandescent light bulb can be used in place of the flashlight bulb.

Skills Practice Lab

DATASHEET A

Star Colors: Red Hot, or Not?

When you look at the night sky, some stars are brighter than others. Some are even different colors. For example, Betelgeuse, a bright star in the constellation Orion, glows red. Sirius, one of the brightest stars in the sky, glows bluish white. Astronomers use color to estimate the temperature of stars. In this activity, you will experiment with a light bulb and some batteries to discover what the color of a glowing object reveals about the temperature of the object.

OBJECTIVES

Discover what the color of a glowing object reveals about the temperature of the object.

Describe how the color and temperature of a star are related.

MATERIALS

- battery, D cell (2)
- battery, D cell, weak
- flashlight bulb
- tape, electrical
- wire, insulated copper, with ends stripped, 20 cm long (2)

SAFETY INFORMATION 

Using Scientific Methods

ASK A QUESTION

1. How are the color and temperature of a star related?

FORM A HYPOTHESIS

2. Remember that a hypothesis is a guess based on facts. Change the question above into a hypothesis.

- The color of the coolest stars is _____________________.

- The color of the hottest stars is _____________________.

TEST THE HYPOTHESIS

3. Every battery has a positive pole (+) and a negative pole (−).
   - Tape one end of an insulated copper wire to the positive pole of the weak D cell battery.
   - Tape one end of the second insulated copper wire to the negative pole of the weak D cell battery.

4. Touch the free end of each wire to the light bulb.
  - Hold one of the wires against the bottom tip of the light bulb.
  - Hold the second wire against the side of the metal part of the bulb.
  - The bulb should light.

5. Look at the filament (wire) inside the light bulb.
  - What color is the filament?

  ____

  - Carefully touch your hand to the bulb. What temperature is the bulb? Is the bulb hot, warm, cool, or cold?

  ____

6. Use one of the two fresh D cells. Repeat steps 3–5.
  - What color is the filament?

  ____

  - What temperature is the bulb?

  ____

7. Use the electrical tape to connect the two fresh D cells.
  - Touch the positive pole of the first cell to the negative pole of the second cell.
  - Tape the two cells together.

8. Using the fresh D cells that are taped together, repeat steps 3–5.
  - What color is the filament?

  ____

  - What temperature is the bulb?

  ____

ANALYZE THE RESULTS

9. Describing Events For each trial, write down the filament colors and bulb temperatures below:

  a. Trial 1: weak D cell: filament color ____________________________

  bulb temperature ____________________________
  - Was the bulb temperature in trial 1 hotter or cooler than in trial 2?

  ____

  - Was the bulb temperature in trial 1 hotter or cooler than in trial 3?

  ____

Star Colors: Red Hot, or Not? *continued*

b. Trial 2: fresh D cell: filament color _______________________

bulb temperature _______________________
- Was the bulb temperature in trial 2 hotter or cooler than in trial 1?

____

- Was the bulb temperature in trial 2 hotter or cooler than in trial 3?

____

c. Trial 3: two fresh D cells: filament color _______________________

bulb temperature _______________________
- Was the bulb temperature in trial 3 hotter or cooler than in trial 1?

____

- Was the bulb temperature in trial 3 hotter or cooler than in trial 2?

____

10. Analyzing Results Star color and surface temperature are a lot like the bulbs you tested.
- What does the color of a star tell you about the star's temperature?

____

____

____

____

11. Classifying Data
- What color are stars that have high surface temperatures?

____

____

- What color are stars that have low surface temperatures?

____

____

DRAW CONCLUSIONS

12. Applying Conclusions Number these stars from highest (1) to lowest (5) surface temperature:

_____ Sirius (bluish white)

_____ Aldebaran (orange)

_____ Procyon (yellow-white)

_____ Capella (yellow)

_____ Betelgeuse (red)

BIG IDEA QUESTION

13. Applying Conclusions What does the yellow color from the sun tell you about the surface temperature of the sun?

____

____

____

____

                                    DATASHEET B

Star Colors: Red Hot, or Not?

When you look at the night sky, some stars are brighter than others. Some are even different colors. For example, Betelgeuse, a bright star in the constellation Orion, glows red. Sirius, one of the brightest stars in the sky, glows bluish white. Astronomers use color to estimate the temperature of stars. In this activity, you will experiment with a light bulb and some batteries to discover what the color of a glowing object reveals about the temperature of the object.

OBJECTIVES

Discover what the color of a glowing object reveals about the temperature of the object.

Describe how the color and temperature of a star are related.

MATERIALS

- battery, D cell (2)
- battery, D cell, weak
- flashlight bulb

- tape, electrical
- wire, insulated copper, with ends stripped, 20 cm long (2)

SAFETY INFORMATION 

Using Scientific Methods

ASK A QUESTION

1. How are the color and temperature of a star related?

FORM A HYPOTHESIS

2. Change the question above into a statement that gives your best estimate about the relationship between the color and temperature of a star.

____

____

TEST THE HYPOTHESIS

3. Tape one end of an insulated copper wire to the positive pole of the weak D cell. Tape one end of the second wire to the negative pole.

4. Touch the free end of each wire to the light bulb. Hold one of the wires against the bottom tip of the light bulb. Hold the second wire against the side of the metal portion of the bulb. The bulb should light.

Star Colors: Red Hot, or Not? *continued*

5. Record the color of the filament in the light bulb. Carefully touch your hand to the bulb. Observe the temperature of the bulb. Record your observations.

____

____

6. Using one of the two fresh D cells, repeat steps 3–5.

____

____

7. Use the electrical tape to connect the two fresh D cells so that the positive pole of the first cell is connected to the negative pole of the second cell.

8. Using the fresh D cells that are taped together, repeat steps 3–5.

____

____

____

ANALYZE THE RESULTS

9. Describing Events What was the color of the filament in each of the three trials? For each trial, compare the temperature of the bulb with the temperature of the bulb in the other two trials.

____

____

____

____

10. Analyzing Results What information does the color of a star tell you about the star?

____

____

____

11. Classifying Data What color are stars that have relatively high surface temperatures? What color are stars that have relatively low surface temperatures?

____

____

____

____

DRAW CONCLUSIONS

12. Applying Conclusions Arrange the following stars in order from highest to lowest surface temperature: Sirius, which is bluish white; Aldebaran, which is orange; Procyon, which is yellow-white; Capella, which is yellow; and Betelgeuse, which is red.

____

____

____

____

BIG IDEA QUESTION

13. Applying Conclusions What does the yellow color of the sun tell you about the surface temperature of the sun?

____

____

____

____

DATASHEET C

Star Colors: Red Hot, or Not?

When you look at the night sky, some stars are brighter than others. Some are even different colors. For example, Betelgeuse, a bright star in the constellation Orion, glows red. Sirius, one of the brightest stars in the sky, glows bluish white. Astronomers use color to estimate the temperature of stars. In this activity, you will experiment with a light bulb and some batteries to discover what the color of a glowing object reveals about the temperature of the object.

OBJECTIVES

Discover what the color of a glowing object reveals about the temperature of the object.

Describe how the color and temperature of a star are related.

MATERIALS

- battery, D cell (2)
- battery, D cell, weak
- flashlight bulb
- tape, electrical
- wire, insulated copper, with ends stripped, 20 cm long (2)

SAFETY INFORMATION ⬥

Using Scientific Methods

ASK A QUESTION

1. How are the color and temperature of a star related?

FORM A HYPOTHESIS

2. Change the question above into a hypothesis.

____

____

TEST THE HYPOTHESIS

3. Tape one end of an insulated copper wire to the positive pole of the weak D cell. Tape one end of the second wire to the negative pole.

4. Touch the free end of each wire to the light bulb. Hold one of the wires against the bottom tip of the light bulb. Hold the second wire against the side of the metal portion of the bulb. The bulb should light.

Star Colors: Red Hot, or Not? *continued*

5. Create a table to record your observations. Record the color of the filament in the light bulb. Carefully touch your hand to the bulb. Observe the temperature of the bulb. Record your observations.

6. Using one of the two fresh D cells, repeat steps 3–5.

7. Use the electrical tape to connect two fresh D cells so that the positive pole of the first cell is connected to the negative pole of the second cell.

8. Using the fresh D cells that are taped together, repeat steps 3–5.

ANALYZE THE RESULTS

9. Describing Events Examine the data in your table. Compare the filament colors and bulb temperatures in the three trials.

____

____

____

10. Analyzing Results What information does the color of a star give you about the star?

____

____

11. Classifying Data What does the color of a star tell you about the relative surface temperature?

____

____

DRAW CONCLUSIONS

12. Applying Conclusions Arrange the following stars in order from highest to lowest surface temperature: Sirius, which is bluish white; Aldebaran, which is orange; Procyon, which is yellow-white; Capella, which is yellow; and Betelgeuse, which is red.

____

____

____

____

Star Colors: Red Hot, or Not? *continued*

BIG IDEA QUESTION

13. Applying Conclusions What does the sun's color tell you about its surface temperature? Which of the stars in question 12 have a similar surface temperature?

____

____

____

____

____

Science Skills Activity **DATASHEET**

Constructing a Line Graph

Teacher Notes

Students may have difficulty choosing which variable to put on which axis and what scales to use for each axis. Distance should go on the horizontal axis (x-axis), with data ranging from 0 to 2,000 million light-years. Lines should be drawn every 250 million or 500 million light-years. Velocity should go on the vertical axis (y-axis), with data ranging from 0 to 70,000 km/s. Lines should be drawn every 5,000 or 10,000 km/s.

The ratio of velocity to distance on the Hubble diagram is Hubble's constant, H_0. Using this data, the constant has units of km/s/Mly (million light-years). Cosmologists usually give Hubble's constant in units of km/s/Mpc (megaparsecs). These units differ by a factor of 3.26. The average of Hubble's constant from this data is 30.8 km/s/Mly • 100 km/s/Mpc. Throughout the latter half of the 20th century, cosmologists debated the value of the Hubble constant, with estimates ranging from 50 to 100 km/s/Mpc. Recent observations from the *Hubble Space Telescope* and the *Wilkinson Microwave Anisotropy Probe* (WMAP) have narrowed the estimates of the Hubble constant to 71•4 km/s/Mpc.

   DATASHEET

Constructing a Line Graph

INVESTIGATION AND EXPERIMENTATION

8.9.c Construct appropriate graphs from data and develop quantitative statements about the relationships between variables.

TUTORIAL

1. Follow these steps to build your graph.
- Draw the axes for your graph on a sheet of graph paper.
- Label the x-axis in terms that describe the types of data. Usually, the independent variable is placed on the x-axis.
- Write the numbers of the scale next to the tick marks of the axis.
- Title the x-axis.
- Determine a scale for the y-axis that includes the range of your data.
- Write the numbers of the scale next to the tick marks of the axis.
- Title the y-axis.
- Write the title of your graph on the top of the graph.

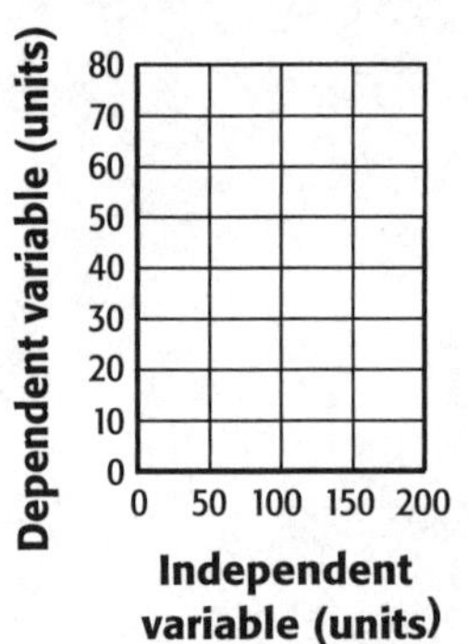

2. Follow these steps to plot your data.
- Plot a series of points by using your data. Each point indicates an (x, y) coordinate on the graph.
- On the graph, draw a line through all of the points that you have plotted.

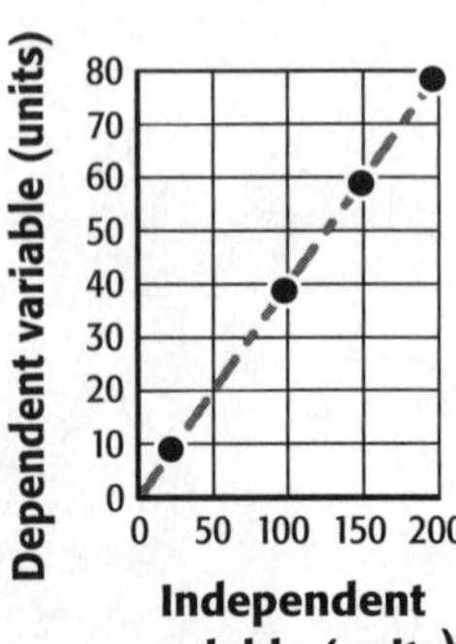

Constructing a Line Graph *continued*

YOU TRY IT!

Procedure

Use the data in **Table 1** to construct a line graph. Then, use the graph to make observations about the relationship between the velocity of a galaxy cluster and the galaxy cluster's distance from Earth. This graph is known as a *Hubble diagram*.

Table 1: Galaxy Clusters

Galaxy Cluster	Distance (in millions of light-years)	Velocity (km/s)
Virgo	39	1,200
Ursa Major	500	15,000
Corona Borealis	700	22,000
Boötes	1,250	39,000
Hydra	1,980	61,000

ANALYSIS

1. Identifying Write a statement that describes the relationship between the velocity of a galaxy cluster and its distance from Earth.

____

____

2. Comparing Compare the ratio of the distance between the galaxy cluster Virgo and Earth to Virgo's velocity with the ratio of the distance between the galaxy cluster Hydra and Earth to Hydra's velocity. What does the comparison in ratios tell you about the line that you will plot on your graph?

____

____

____

3. Evaluating If a galaxy cluster is more than 1,980 million light-years from Earth, at what velocity is that galaxy cluster moving?

____

____

____

____

Answer Key

Directed Reading A

SECTION: STARS
1. B
2. C
3. A
4. C
5. A
6. D
7. continuous spectrum
8. emission lines
9. D
10. B
11. A
12. B
13. A
14. C
15. A
16. D
17. C
18. B
19. A
20. A
21. B
22. B
23. apparent magnitude
24. absolute magnitude
25. light-year
26. parallax
27. A
28. C

SECTION: THE LIFE CYCLE OF STARS
1. B
2. D
3. C
4. D
5. A
6. B
7. D
8. B
9. A
10. B
11. D
12. C
13. D
14. C
15. B
16. A
17. supernova
18. neutron star
19. pulsar
20. black hole

SECTION: GALAXIES
1. D
2. C
3. D
4. A
5. B
6. B
7. C
8. D
9. B
10. B
11. B
12. B

SECTION: FORMATION OF THE UNIVERSE
1. D
2. B
3. A
4. D
5. B
6. A
7. C
8. C
9. B
10. A
11. C
12. C
13. C
14. C
15. B
16. D

Directed Reading B

SECTION: STARS
1. Answers may vary. Sample answer: A star is a huge, hot, bright ball of gas.
2. starlight
3. A
4. Stars that differ in color also differ in temperature.
5. D
6. A
7. D
8. D
9. A

10. Answers may vary. Sample answer: The cooler atmosphere of a star absorbs colors of light instead of emitting them.

11. Answers may vary. Sample answer: An absorption spectrum is produced when light from a hot solid or dense gas passes through a less dense, cooler gas. The cooler gas absorbs portions of the spectrum.

12. The black lines of a star's spectrum represent portions of the spectrum that are absorbed by the star's atmosphere.

13. The pattern of lines in a star's absorption spectrum is unique to that star and to the stage that the star occupies in its life cycle.

14. A star is a mixture of elements, and all of the different lines for a star's elements appear together in its spectrum.

15. hydrogen and helium

16. carbon, nitrogen, oxygen

17. A

18. B

19. D

20. first-magnitude

21. positive numbers

22. negative numbers

23. apparent magnitude

24. absolute magnitude

25. Answers may vary. Sample answer: Although the absolute magnitude of the sun is +4.8, which is ordinary for a star, the sun's apparent magnitude is -26.8 because it is so close to Earth.

26. C

27. light-year

28. parallax

29. Answers may vary. Sample answer: During each different season of the year, Earth faces a different part of the sky at night.

30. Earth's rotation

31. Answers may vary. Sample answer: Even though each star is moving in space, their actual motions are difficult to see because stars are so distant.

SECTION: THE LIFE CYCLE OF STARS

1. Stars can be classified by mass, size, brightness, color, temperature, composition, and age.

2. Answers may vary. Sample answer: The classification of a star changes as its properties change.

3. D

4. C

5. A

6. B

7. A

8. Temperature appears along the bottom of the H-R diagram (along the horizontal axis).

9. Absolute magnitude appears along the left side of the H-R diagram (along the vertical axis).

10. in the main sequence

11. C

12. A

13. B

14. D

15. As they age, main-sequence stars move up and to the right to become giants or supergiants, then move down and to the left to become white dwarfs.

16. C

17. A

18. D

19. B

20. Black holes do not give off light. Gas and dust from nearby stars may spiral into the black hole and give off X rays that astronomers can detect.

SECTION: GALAXIES

1. galaxy

2. E

3. S

4. I

5. S

6. E

7. I

8. S

9. D

10. A

11. B

12. quasars

13. It takes time for light to travel through space, so looking through a telescope is like looking back through time. The farther out one looks, the farther back in time one sees.

14. Scientists study distant galaxies to learn what early galaxies looked like. This gives them information about how galaxies change over time and what may have caused them to form.

SECTION: FORMATION OF THE UNIVERSE

1. cosmology
2. A
3. D
4. the big bang theory
5. Answers may vary. Sample answers: between 13 and 15 billion years ago; about 14 billion years ago; about 13.7 billion years ago
6. Answers may vary. Sample answer: Cosmic background radiation is energy left over from the original big bang explosion that was distributed in every direction as the universe expanded.
7. Answers may vary. Sample answer: After the big bang, gravitational attraction caused matter to form galaxies, and the attraction between galaxies caused galaxies to cluster.
8. Earth is part of the solar system; the solar system is part of the Milky Way galaxy, and the Milky Way is part of a galaxy cluster
9. D
10. The oldest white dwarfs are between 12 and 13 billion years old. It took about one billion years after the big bang for the first white dwarfs to form, so the universe must be approximately 14 billion years old.
11. C
12. D
13. C
14. Dark energy seems to be accelerating the expansion of the universe, counteracting the effect of gravity.
15. Answers may vary. Sample answer: If the expansion rate of the universe continues to grow, stars will age and die, the universe will become cold and dark, but the universe will continue to expand forever.

Vocabulary and Section Summary A

SECTION: STARS

1. spectrum: the band of colors produced when white light passes through a prism
2. apparent magnitude: the brightness of a star as seen from Earth
3. absolute magnitude: the brightness that a star would have at a distance of 32.6 light-years from Earth
4. light-year: the distance that light travels in one year; about 9.46 trillion kilometers
5. parallax: an apparent shift in the position of an object when viewed from different locations

SECTION: THE LIFE CYCLE OF STARS

1. main sequence: the location on the H-R diagram where most stars lie; it has a diagonal pattern from the lower right (low temperature and luminosity) to the upper left (high temperature and luminosity)
2. H-R diagram: Hertzsprung-Russell diagram, a graph that shows the relationship between a star's surface temperature and absolute magnitude
3. supernova: a gigantic explosion in which a massive star collapses and throws its outer layers into space

SECTION: GALAXIES

1. galaxy: a collection of stars, dust, and gas bound together by gravity
2. nebula: a large cloud of gas and dust in interstellar space; a region in space where stars are born

SECTION: FORMATION OF THE UNIVERSE

1. big bang theory: the theory that all matter and energy in the universe was compressed into an extremely small volume that exploded 13 billion to 15 billion years ago and began expanding in all directions

Vocabulary and Section Summary B

SECTION: STARS

1. spectrum
2. apparent magnitude
3. absolute magnitude
4. light-year
5. parallax

SECTION: THE LIFE CYCLE OF STARS

1. main sequence
2. H-R diagram
3. supernova
4. neutron star
5. pulsar
6. black hole

T	M	W	H	U	B	E	L	Q	Y	N	I	T	O	T
P	H	B	G	P	C	N	K	O	F	E	K	J	Z	Z
E	B	K	Q	W	N	N	V	V	D	U	M	Y	Q	L
N	E	H	V	S	P	B	R	X	B	T	H	F	Q	Y
T	Y	I	M	R	U	Q	Z	O	I	R	F	D	W	O
M	N	B	A	B	L	A	C	K	H	O	L	E	F	M
K	P	H	D	U	S	U	P	E	R	N	O	V	A	I
M	M	G	U	Y	A	S	Q	J	I	S	O	R	J	W
W	X	S	P	S	R	N	F	B	N	T	G	Q	H	C
E	C	N	E	U	Q	E	S	N	I	A	M	Q	Q	M
L	D	P	J	N	Y	P	M	D	I	R	M	B	C	U
J	A	L	A	C	W	G	Z	D	E	T	J	L	N	X
U	V	W	T	T	P	U	R	M	H	A	Q	U	X	X
A	W	M	W	C	T	H	P	R	U	T	T	V	M	H
K	O	S	V	P	W	J	F	J	Z	X	E	D	A	C

SECTION: GALAXIES

Across

    5. spiral galaxy

    7. nebula

Down

    1. open cluster

    2. globular cluster

    3. Magellanic Clouds

    4. Milky Way

    6. galaxy

SECTION: FORMATION OF THE UNIVERSE

    1. white dwarfs

    2. dark energy

    3. cosmology

    4. cosmic background radiation

    5. dark matter

    6. big bang theory

    7. star cluster

Reinforcement

DIAGRAMMING THE STARS

1.-6.

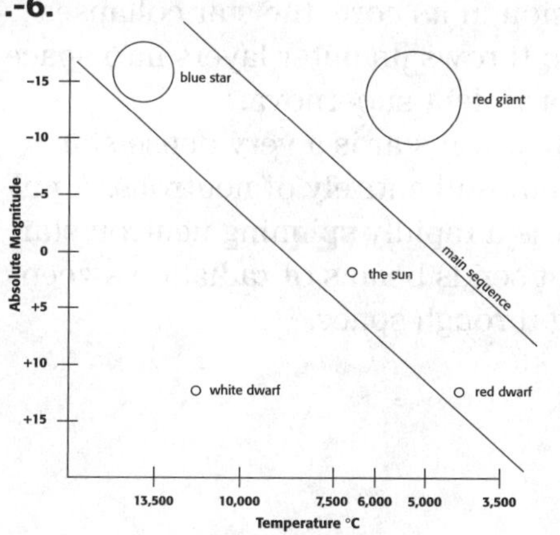

7. The blue star, the sun, and the red dwarf star are included in the main sequence.

8. Answers may vary. Sample answer: I think that it is a white dwarf star.

Critical Thinking

1. Answers may vary. Sample answer: The branches might represent galaxies.

2. Answers may vary. Sample answer: The leaves could represent stars.

3. No; it is a mythical explanation of how the universe formed. It relies on metaphors and supernatural animals, and it does not use the scientific method.

4. Answers may vary. Sample answer: Yes, it does resemble the big bang theory in some ways. Like the big bang theory, this explanation states that the universe began as a single point that contained all the matter in the universe. Also like the big bang theory, this explanation states that at one point all of this material exploded outward in every direction from a central location.

5. Answers may vary. Sample answer: The big bang theory does not explain why an explosion occurred, while this explanation does.

6. Answers may vary. Sample answers: Yes; I think science will eventually solve the mysteries of the universe because scientific tools and technology are becoming increasingly advanced. No, I don't think that science will solve the mysteries of the universe because the structures, forces, and number and type of objects in the universe are too complex to ever completely be understood using the scientific method.

SciLinks Activity

Answers may vary. Sample answers:

Spiral galaxies have a bulge at the center and spiral arms. The spiral arms are made up of gas, dust, and new stars that have formed in these denser regions of gas and dust.

Elliptical galaxies may look like spheres, or be more stretched out. Some are cucumber shaped, with the round end facing our galaxy. Elliptical galaxies usually have very bright centers and very little dust and gas. Elliptical galaxies contain mostly old stars. Some elliptical galaxies are huge and are called giant elliptical galaxies. Other elliptical galaxies are much smaller and are called dwarf elliptical galaxies.

Irregular galaxies are galaxies that don't fit into any other class. Their shape is irregular.

Section Review

SECTION: STARS

1. Sample answer: Apparent magnitude is how bright a star looks from Earth, and absolute magnitude is how bright a star would look at a distance of 32.6 light-years from Earth.
2. Sample answer: Spectrum is the band of colors produced when light passes through a prism. A light-year is the distance that light travels in one year. Parallax is the apparent shift in the position of an object when the object is viewed from different locations.
3. Stars that are blue have the highest temperatures. Stars that are red have the lowest temperatures.
4. Each element has a unique absorption spectrum. Astronomers can find the composition of a star by looking for the absorption spectra of individual elements in the spectrum of the star.
5. Hydrogen and helium are two elements most commonly found in stars.
6. Apparent magnitude is the brightness of a star as it appears from Earth. Absolute magnitude is the actual brightness of the star as it would appear at a distance of 32.6 light-years from Earth.
7. Astronomers use the unit light-years to measure the distance from Earth to stars. One light-year equals about 9.46 trillion kilometers.

8. Because the distances from Earth to stars are so great, it is easier to express these distances by using light-years.
9. Stars appear to move in the sky because of Earth's rotation on its axis.
10. A continuous spectrum shows all colors that are emitted by a hot object. An absorption spectrum shows where certain colors are absorbed instead of emitted. Astronomers can find the composition of a star by finding the unique absorption spectra of different elements in the spectrum of the star.
11. The absolute magnitude of the sun is relatively dim compared to other stars, but its apparent magnitude is extremely bright because it is much closer to Earth than other stars are.
12. If a star displayed a large parallax, you could say that the star is close to Earth.
13. $90 \text{ m} \div 10 \text{ m} = 9$; $9 \times 9 = 81$ times as bright
14. approximately 6 hours

SECTION: THE LIFE CYCLE OF STARS

1. Sample answer: Stars on the main sequence form a diagonal band on the H-R diagram.
2. New stars form when gravity pulls gas and dust together into a sphere. When the sphere is dense enough, it starts to convert hydrogen into helium through nuclear fusion in its core.
3. main-sequence star, red giant, white dwarf
4. In main-sequence stars, the hotter a star is, the brighter it is, and the cooler a star is, the dimmer it is.
5. When a massive star stops nuclear fusion in its core, the star collapses and throws its outer layers into space in a violent supernova.
6. A neutron star is a very dense star composed entirely of neutrons. A pulsar is a rapidly spinning neutron star that sends beams of radiation sweeping through space.

7. If the center of a collapsed star has a mass several times the mass of the sun, it may collapse further, crushing the dense center of the star to form a black hole.

8. By plotting stars according to brightness and temperature, the H-R diagram shows that stars exist in distinct groups corresponding to life stages. Stars move from one group to another on the diagram as they pass through the stages in their life cycles.

9. Massive stars have shorter life spans than other stars do because massive stars burn up their supply of hydrogen more quickly.

10. After a star explodes as a supernova, it may collapse to form a neutron star, a pulsar, or a black hole.

11. The astronomer can look for X rays, which would be emitted if gas and dust from a nearby star were spiraling into a black hole.

SECTION: GALAXIES

1. Galaxies are made of stars and clouds of gas and dust called *nebulas*.

2. Spiral galaxies have a bulge in the center and spiral arms. They may contain billions of stars. New stars form in dense regions of gas and dust in the spiral arms. Elliptical galaxies are round or slightly flattened. They may contain trillions of stars. No new stars are forming in elliptical galaxies. Irregular galaxies have no definite shape. They may contain millions to billions of stars. New stars form slowly in irregular galaxies.

3. The Milky Way galaxy has a bulge in the center and spiral arms. It contains about 200 billion stars. The sun is located about two-thirds of the way between the center and the edge of the galaxy. New stars form in denser regions of gas and dust in the spiral arms of the galaxy.

4. A nebula is a large cloud of dust and gas where stars are born. Star clusters are groups of stars that formed at the same time from the same nebula.

5. Globular clusters are most likely to be found in the spherical halo around spiral galaxies or near giant elliptical galaxies.

6. Scientists see what early galaxies looked like by observing galaxies that are extremely far away. Because light takes time to travel through space, looking at distant galaxies is like looking far back in time.

7. A globular cluster is a highly concentrated group of up to 1 million stars. A globular cluster is located either in the spherical halo that surrounds spiral galaxies or close to some elliptical galaxies. An open cluster is a group of 100 to 1,000 stars that is located close together relative to other stars. An open cluster is located along the disk of a spiral galaxy.

8. Looking through a telescope is like looking back in time because light takes such long periods of time to travel through space.

9. Scientists can compare galaxies that are extremely far away in space with less-distant, younger galaxies to see how galaxies change over time.

SECTION: FORMATION OF THE UNIVERSE

1. The big-bang theory states that the universe began with a tremendous explosion. About 14 billion years ago, all of the contents of the universe were compressed into a very small volume. As the universe rapidly expanded from that state, the initial elements, natural forces, and seeds of galaxy formation were formed.

2. The force of gravity came into being along with the rest of the universe during the big bang. While the universe is expanding, gravity counteracts the expansion and causes matter to be pulled together into galaxies, solar systems, stars, and planets.

3. According to the big-bang theory, energy from the big bang was distributed in every direction. The cosmic background radiation is found in every direction in space, so it may be energy left over from the big bang.

4. The universe has a structure of galaxies and star systems that is loosely repeated over and over again, and every object in the universe is part of a larger system.

5. If the universe expands forever, it will probably become entirely dark and cold.

6. White dwarfs are at the end of the life cycle of long-lived stars, so the oldest stars in the universe are most likely white dwarfs.

7. 4 cm/s, 12 cm/s, 8 cm/s

Chapter Review

1. C
2. apparent magnitude
3. a light-year
4. H-R diagram
5. B
6 A
7. C
8. D
9. B
10. Scientists classify stars by temperature, brightness, composition, and age.
11. Scientists use light-years to express distances from Earth to the stars because the distances are very great.
12. A star is born, becomes a main-sequence star, uses up its hydrogen, and becomes a red giant or supergiant. Very massive stars may end their lives in supernovas, then become neutron stars or black holes. Less massive stars may end up as white dwarfs.
13. When a star is on the main sequence, nuclear fusion of hydrogen into helium generates energy. This energy balances the inward pull of gravity.
14. Spiral galaxies have a bulge in the center and spiral arms. They may contain billions of stars. New stars form in dense regions of gas and dust in the spiral arms. Elliptical galaxies are round or slightly flattened. They may contain trillions of stars. No new stars are forming in elliptical galaxies. Irregular galaxies have no definite shape. They may contain millions to billions of stars. New stars form slowly in irregular galaxies.

15. The Milky Way galaxy has a bulge in the center and spiral arms. It contains about 200 billion stars. The sun is about two-thirds of the way between the center of the galaxy and the edge of the galaxy. New stars form in denser regions of gas and dust in the galaxy.

16. According to the big-bang theory, energy from the big bang was distributed in every direction. The cosmic microwave background radiation is found in every direction in space, so it may be energy left over from the big bang.

17. Dark matter is matter that does not give off light but that may be detected indirectly by its gravity. Dark energy is unknown energy that is thought to be accelerating the expansion of the universe.

18. Answers may vary.

19. An answer to this exercise can be found at the end of the Teacher Edition.

20. If a star displayed a large parallax, you could say the star is close to Earth.

21. Main-sequence stars are stars that are generating energy by fusing hydrogen into helium in their cores. Giant or supergiant stars have run out of hydrogen. Their centers have shrunk inward, and their outer layers have expanded outward. White dwarfs are the leftover centers of old stars. White dwarfs are small and hot.

22. It is most likely an elliptical galaxy. New stars are no longer created in elliptical galaxies, so most stars in an elliptical galaxy are in the later stages of their lives.

23. All the contents of the universe were contained in a very small volume. The universe began about 14 billion years ago when it expanded rapidly in all directions. Cosmic microwave background radiation is energy left over from the "big bang" that started this expansion.

24. A star with half the mass of the sun would live longer because it would use up its hydrogen more slowly than a more massive star would.

25. A main-sequence star that has a mass about 1.5 times the mass of the sun would live about 5 billion years.

26. The galaxy is about 90 million light-years from Earth.

27. The galaxy is moving away from Earth at about 15,000 km/s.

28. $15 - -5 = 20$; $20 \div 5 = 4$; $100^4 = 100 \times 100 \times 100 \times 100 = 100$ million; The star that has a magnitude of -5 is 100 million times brighter than the star that has a magnitude of $+15$.

29. More stars will end their lives as white dwarfs.

30. Dark energy seems to be counteracting the force of gravity and accelerating the expansion of the universe.

Section Quizzes

SECTION: STARS

1. B	**6.** E
2. C	**7.** D
3. B	**8.** C
4. D	**9.** A
5. D	**10.** B

SECTION: THE LIFE CYCLE OF STARS

1. D	**6.** D
2. C	**7.** B
3. A	**8.** E
4. D	**9.** A
5. B	**10.** C

SECTION: GALAXIES

1. D	**6.** B
2. B	**7.** D
3. A	**8.** E
4. C	**9.** C
5. B	**10.** A

SECTION: FORMATION OF THE UNIVERSE

1. A	**5.** C
2. B	**6.** A
3. A	**7.** C
4. B	

Chapter Test A

1. A
2. D
3. A
4. C
5. A
6. D

7. B
8. D
9. B
10. B
11. A
12. A
13. B
14. A
15. C
16. A
17. C
18. B
19. light-year
20. main sequence
21. apparent magnitude
22. parallax

Chapter Test B

1. D
2. A
3. D
4. A
5. B
6. D
7. D
8. C
9. D
10. B
11. C
12. C
13. C
14. C
15. A
16. D
17. B
18. D
19. A
20. D
21. D
22. C
23. A
24. E
25. B

Chapter Test C

1. absolute magnitude
2. gravity
3. cosmic background radiation
4. black hole
5. globular cluster
6. spectrum
7. A

8. A

9. B

10. B

11. D

12. D

13. B

14. Answers may vary. Sample answer: Spiral galaxies have a bulge in the center and spiral arms made up of gas, dust, and new stars. Elliptical galaxies are round galaxies without spiral arms that have stopped making new stars. Irregular galaxies have no definite shape and form stars very slowly.

15. Sample answer: The absorption spectra of stars can be compared with the emission spectra produced by elements. If the spectral lines of the star match certain elements, then it can be concluded that these elements are in the star or in its atmosphere.

16. Answers may vary. Sample answer: If the expansion rate of the universe continues to increase, the expansion of the universe will continue forever, even though stars will age and die, and after billions of years the universe will become cold and dark.

17. Sample answer: The stars appear to move around the North Star because Earth is turning on its axis, which is pointed toward the North Star.

18. 1.5×10^{11} m $\div$ 3 $\times$ 108 m/s $\div$ 60 s/min = 8 $\times 10^0$ minutes = 8 minutes. It would take about 8 minutes for the last ray of sunlight to reach Earth.

19. The temperature of the star is just above 10,000°C, and it has an absolute magnitude of about +10.5.

20. a. galaxy clusters, **b.** galaxies, **c.** nebulas, **d.** star clusters, **e.** globular clusters, **f.** planets

Performance-Based Assessment

1. 50 ly $\div$ 10 ly / cm = 5 cm; 60 ly $\div$ 10 ly/ cm = 6 cm; 75 ly $\div$ 10 ly / cm = 7.5 cm 90 ly $\div$ 10 ly / cm = 9 cm; 100 ly $\div$ 10 ly / cm = 10 cm; 200 ly $\div$ 10 ly / cm = 20 cm

5. Mizar looks closest to Alkaid when viewed from Earth.

6. Distance from Earth and actual brightness affect a star's apparent magnitude.

7. Megrez and Alioth would look equally bright because they are the closest distance to Earth.

8. A light-year is the distance light travels in one year.

9. Sample answer: The sun is 8 light-minutes from Earth.

10. No. Even though Mizar looks like the closest star to Alkaid, Dubhe may actually be closer.

Explore Activity

DATASHEET A

5. As the rubber band was stretched, each galaxy moved away from neighboring galaxies.

6. When I stretch the rubber band, E moves 4 cm away from A at the same time as B moves only 1 cm away. So, E appears to move away from A 4 times faster than B does.

DATASHEET B

5. As the rubber band was stretched, each galaxy moved away from neighboring galaxies.

6. When I stretch the rubber band, E moves 4 cm away from A at the same time as B moves only 1 cm away. So, E appears to move away from A 4 times faster than B does.

DATASHEET C

5. As the rubber band was stretched, each galaxy moved away from neighboring galaxies.

6. When I stretch the rubber band, E moves 4 cm away from A at the same time as B moves only 1 cm away. So, E appears to move away from A 4 times faster than B does.

Quick Lab: Demonstrating Parallax

DATASHEET A

3. The object I covered in step 2 seems to have moved to the right of my thumb when I close my left eye.

4. When my thumb is close to my eye, the object seems to have moved more to the right than when my thumb was farther away.

5. The closer an object is, the more it appears to shift position.

DATASHEET B

3. Your thumb appears to move to the left or right of the object, depending on which eye is closed.

5. The closer an object is, the more it appears to shift position.

DATASHEET C

3. Your thumb appears to move to the left or right of the object, depending on which eye is closed.

4. Your thumb appears to move farther to the left or right of the object, depending on which eye is closed.

5. The closer an object is, the more it appears to shift position.

Quick Lab: Making a Star Movie

DATASHEET A

5. The main-sequence stage needs the most pages, because a star stays longest in that stage.

6. A star is usually largest and brightest during the giant or supergiant stage. If the student has drawn a supernova, that stage may be identified as the brightest. A star is smallest and dimmest during the last stage of its life, as a white dwarf, a neutron star, or a black hole.

DATASHEET B

5. The main sequence stage needs the most pages, because a star stays longest in that stage.

6. A star is typically largest and brightest during the giant or supergiant stage, but it may be much brighter for a short time as a supernova. A star is smallest and dimmest during the last stage of its life, whether it is a white dwarf, a neutron star, or a black hole.

DATASHEET C

5. The main-sequence stage needs the most pages, because a star stays longest in that stage.

6. A star is typically largest and brightest during the giant or supergiant stage, but it may be much brighter for a short time as a supernova. A star is smallest and dimmest during the last stage of its life, whether it is a white dwarf, a neutron star, or a black hole.

Quick Lab: Modeling Galaxies

DATASHEET A

4. a spiral pattern

5. a spiral galaxy

DATASHEET B

4. a spiral pattern

5. a spiral galaxy

DATASHEET C

4. The glitter forms a spiral pattern.

5. a spiral galaxy

Quick Lab: The Expanding Universe

DATASHEET A

6. A and C

7. C

DATASHEET B

6. A and C

7. The farther galaxies are from Earth, the greater the rate at which they are moving away from Earth is.

DATASHEET C

6. A and C

7. The farther galaxies are from Earth, the greater the rate at which they are moving away from Earth is.

Chapter Lab

DATASHEET A

5. The filament is dull red. The bulb is warm.

6. The filament is bright red or orange. The bulb is warm or hot.

8. The filament is white or almost white. The bulb is hot.

9. The bulb in trial 1 is cooler than in trial 2 or 3. The bulb in trial 2 is hotter than in trial 1 and cooler than in trial 3. The bulb in trial 3 is hotter than in trial 1 or trial 2.

10. The cooler filament emitted red light. As the bulb became hotter, its color gradually changed from red to orange to white. Like the color emitted by the filament, the color of the light emitted from stars helps scientists determine the surface temperature of stars.

11. White or blue stars have a relatively high surface temperature. Red or orange stars have a relatively low surface temperature.

12. The order of the stars from highest to lowest surface temperature is as follows: 1. Sirius, 2. Procyon, 3. Capella, 4. Aldebaran, and 5. Betelgeuse.

13. The yellow color of the sun indicates that the surface of the sun is relatively cool.

DATASHEET B

9. With the weaker cell, the filament should glow with a dull red color. The filament glows bright red or orange with the stronger cell. With two fresh D cells, the filament becomes almost white. The temperature increases as more cells, or fresh cells, are used.

10. Cooler objects emit red light. As an object becomes hotter, its color gradually changes from red to orange to white. Because the same principle applies to stars, the color of the light emitted from stars helps scientists determine the surface temperature of stars.

11. Stars that have relatively high surface temperatures are white or blue. Stars that have relatively low surface temperatures are red or orange.

12. The order of the stars from highest to lowest surface temperature is as follows: Sirius, Procyon, Capella, Aldebaran, and Betelgeuse.

13. The yellow color of the sun indicates that the surface of the sun is relatively cool.

DATASHEET C

9. With the weaker cell, the filament should glow with a dull red color. The filament glows bright red or orange with the stronger cell. With two fresh D cells, the filament becomes almost white. The temperature increases as more cells—or fresh cells—are used.

10. Cooler objects emit red light. As an object becomes hotter, its color gradually changes from red to orange to white. Therefore, the color of the light emitted from stars helps scientists determine the surface temperature of stars.

11. Stars that have relatively high surface temperatures are white or blue. Stars that have relatively low surface temperatures are red or orange.

12. The order of the stars from highest to lowest surface temperature is as follows: Sirius, Procyon, Capella, Aldebaran, and Betelgeuse.

13. The yellow color of the sun indicates that the surface of the sun is relatively cool. Capella and Procyon have a similar surface temperature.

Science Skills Activity

DATASHEET

1. Sample answer: Galaxy clusters at greater distances from Earth are moving at a greater velocity.

2. The similarity of these ratios suggests that the line on the graph will be close to a straight line.

3. A galaxy at a distance more than 1,980 light-years from Earth would be moving at a velocity greater than 61,000 km/s.